Vetter Theory

Chapter 1

1. Introduction: Challenging Conventional Physics

Why We Need a New Approach

Science is built on questions, and some of the biggest questions remain unanswered. What exactly is dark matter, the unseen substance that makes up most of the universe's mass? What is dark energy, the mysterious force driving the universe's expansion? And how do we reconcile the nature of gravity with our understanding of quantum mechanics?
These mysteries persist because our current theories—especially those grounded in relativity and quantum mechanics—don't offer complete answers. Relativity, which relies on the curvature of spacetime, has been a useful tool, but it is complex and struggles to unify with the quantum world. Similarly, while quantum mechanics has successfully described microscopic particles, it hasn't bridged the gap to explain cosmic-scale phenomena.
This book presents an alternative path. The Vetter Theory is an approach that doesn't rely on the warping of spacetime or complex mathematical frameworks but instead offers a simpler, more direct explanation for gravity, matter, and the structure of the universe. This theory grew out of frustration with traditional explanations and a drive to find answers that are grounded in observable, particle-based phenomena. As we journey through the following chapters, you'll discover a view of the universe where simplicity replaces complexity and where some of the most puzzling aspects of the cosmos gain new clarity.

Dispelling the Concept of Curved Spacetime

One of the first things we'll address is the role of relativity in modern physics. According to relativity, gravity arises from the curvature of spacetime—space and time blended together and bent by mass. But in the Vetter Theory, space and time are distinct and uncurved, and gravity does not result from bending anything.
Imagine a universe where space and time are independent and simply coexist. Gravity, in this view, is not a curvature or warping but a force derived from particle interactions at the smallest scales. In this book, we'll look at gravity not as an effect of mass bending spacetime but as something arising directly from the properties of particles, specifically the proton. This shift in perspective may initially feel unfamiliar, but it leads to a surprisingly intuitive understanding of the forces we observe.

Core Principles of the Vetter Theory

To understand this theory, we'll begin with some key ideas:

- **The Proton as the Source of All Forces**: In the Vetter Theory, the proton isn't just a particle—it's the source of all fundamental forces, with the strong force manifesting itself in ways that create gravity, electromagnetism, and the weak nuclear force. This means the proton is at the heart of every force that governs matter and energy in the universe.
- **Quarks as Dark Matter**: Quarks are often confined within protons and neutrons, but in this theory, free quarks permeate space itself, much like nitrogen fills Earth's atmosphere. These quarks, invisible to us because they don't have electrons, form the elusive dark matter that holds galaxies together.
- **Continuous Creation and Destruction of Matter**: In the Vetter Theory, matter isn't static; it's constantly being created and annihilated. Black holes play a key role, collecting protons while ejecting electrons back into space. These electrons interact with quarks in space, influencing the "pressure" of quarks, which we perceive as dark energy.

This continuous cycle of matter recycling keeps the universe dynamic and self-regulating, unlike the static models suggested by some conventional theories.

A New Perspective on Black Holes and Gravitational Lensing

What exactly is a black hole? The Vetter Theory sees black holes not as bottomless pits or warps in spacetime but as proton collectors. Inside a black hole, protons accumulate in vast numbers while electrons are ejected at near-light speeds. As these high-speed electrons move through space, they influence light in ways that create the effect we call gravitational lensing. Unlike the traditional view, which explains lensing as the result of spacetime bending, here it's a direct interaction between light and high-energy particles.

From Microscopic Interactions to Cosmic Scale Phenomena

What if the forces that govern particles are also responsible for the largest structures in the universe? The Vetter Theory aims to bridge the gap between micro- and macroscales, offering a unified explanation that connects the behavior of particles with the behavior of galaxies and beyond.

By understanding gravity, dark matter, and dark energy as effects of particle interactions, this theory avoids the need for abstract and complex concepts, offering instead a straightforward approach to understanding cosmic forces. It's a return to simplicity that aligns closely with observable phenomena.

What You'll Discover in This Book

In the coming chapters, we'll break down each element of this theory, from the fundamental role of the proton to the nature of black holes and dark matter. You'll see how each piece fits together to create a cohesive, intuitive view of the universe that's free from the constraints of relativity.

This book is an invitation to explore a new framework for understanding the universe, one that moves beyond convention and encourages a fresh perspective. Whether you're a scientist, a curious thinker, or someone with a passion for understanding how the universe works, you'll find that the Vetter Theory offers a radical yet simple approach to answering the deepest questions of existence.

1.2 Dispelling the Concept of Curved Spacetime

The Limitations of Curved Spacetime

For over a century, the idea that gravity results from the curvature of spacetime has shaped our understanding of the universe. General relativity, introduced by Einstein in 1915, suggests that mass bends the fabric of spacetime, creating "curved" paths that objects follow. This model has been successful in predicting certain phenomena, from planetary orbits to gravitational lensing.

However, the Vetter Theory takes a fundamentally different approach. In this view, space and time are distinct and uncurved, and gravity doesn't result from any kind of bending. Instead, gravity is seen as an emergent effect of interactions at the particle level, primarily influenced by the behavior of protons, electrons, and quarks.

Why Abandon Curved Spacetime?

Curved spacetime, while elegant in theory, becomes unwieldy when it encounters the quantum world. Despite extensive research, general relativity remains incompatible with quantum mechanics, the science of particles on microscopic scales. This disconnect has driven scientists to pursue a "theory of everything" that can unify the two domains, yet the reconciliation remains elusive.

In the Vetter Theory, the incompatibility of relativity with the quantum world suggests that perhaps the concept of spacetime curvature is fundamentally flawed. By rejecting the notion of curved spacetime, we can instead seek a model where gravity is not a result of warped geometry but of direct, observable interactions among particles. Gravity becomes a particle-driven force, simpler to understand and intuitively connected to the structures and phenomena we observe in the universe.

Rethinking Gravity Without Spacetime Curvature

So, what replaces the curvature of spacetime? In the Vetter Theory, gravity is a manifestation of the strong force, which typically operates at the atomic scale to bind protons and neutrons within atomic nuclei. However, in this new model, the strong force behaves differently depending on the conditions and density of matter.

When enough matter coalesces, as in massive stars or black holes, the strong force extends beyond its usual range, creating what we perceive as gravitational attraction. This effect emerges not from any curvature in space but from the "saturation" of the strong force—a sort of threshold effect where the force radiates outward and influences surrounding matter. This new approach ties gravity directly to matter density and the behavior of protons, rather than abstract geometry.

Space and Time as Separate Entities

In this theory, space and time are not a single, interwoven fabric but are instead independent dimensions. Time progresses uniformly and is unaffected by the presence of mass. Space, similarly, remains unaffected by time or by the presence of massive objects. This separation offers a simplified, stable view of the universe that avoids the complexities and paradoxes of spacetime curvature.

This view also has practical implications. Concepts like time dilation and length contraction, which are central to relativity, no longer apply. Rather than experiencing time differently based on gravitational fields or velocity, time is absolute, progressing uniformly for all objects. Space, likewise, retains a consistent, unbending nature, unaffected by the proximity of mass.

Gravity as a Particle Interaction, Not a Field Effect
In the Vetter Theory, gravity isn't a "field" that warps space. Instead, it is a force that originates from the behavior of particles, specifically protons and their interactions with quarks and electrons. Here's how it works:

- **Protons as Gravity's Source**: As protons accumulate in massive quantities, their collective interactions produce an attractive force that extends outward, affecting nearby particles.
- **Electron Influence**: Electrons, which are naturally bound to protons in stable atoms, play a secondary role by spacing out protons and influencing the intensity of the force.
- **Beyond the Strong Force's Traditional Limits**: As mass accumulates, the strong force intensifies and extends beyond atomic scales, creating a gravitational effect that binds matter on a large scale.

This reimagined gravity is an emergent force—produced by the density and arrangement of protons rather than by a mass-induced warping of space. Objects experience gravitational attraction not because they are following a curved path in spacetime but because they are responding to the direct influence of matter in their vicinity.

Implications for Astrophysics and Cosmology
With gravity redefined, the Vetter Theory offers new explanations for phenomena traditionally attributed to spacetime curvature. For example:

- **Gravitational Lensing**: In this framework, light bends near massive objects not because of a warp in space but due to the high-speed electrons ejected from black holes, which affect light's path through particle interactions.
- **Black Hole Gravity**: Black holes are treated not as singularities in spacetime but as dense proton accumulators. Their gravity arises from the saturation of protons and the extended influence of the strong force, with electron ejections contributing to lensing effects.

A Simpler, Unified Model
By removing the concept of curved spacetime, the Vetter Theory presents a simpler, more unified model. Gravity, electromagnetism, and other forces are all seen as manifestations of particle interactions rather than abstract fields. This not only resolves the dissonance between quantum mechanics and gravity but also brings a level of clarity to how forces operate universally.

Rejecting spacetime curvature may seem radical, but as we'll see, this approach eliminates many of the mathematical complexities and paradoxes that arise in relativity, such as singularities and time distortions. By focusing instead on direct interactions among particles, we gain a cleaner, more intuitive framework for understanding gravity, one that scales seamlessly from atoms to galaxies.

1.3 Core Principles of the Vetter Theory

Rethinking Fundamental Forces: The Proton as the Source of All Interactions

In the Vetter Theory, the proton is more than just a component of atomic nuclei; it is the key to understanding all the forces that govern matter and energy. Traditional physics identifies four fundamental forces—gravity, electromagnetism, the strong nuclear force, and the weak nuclear force. However, in this theory, the proton itself is the origin of all these forces, with the strong force manifesting in different ways to produce the effects we currently understand as separate forces.

- **The Strong Force as the Root of All Forces**: Here, the strong nuclear force, usually confined to holding quarks together within protons and neutrons, is reimagined as a versatile force that manifests differently depending on the environment. Under certain conditions, it appears as gravity, while in others, it produces the effects attributed to electromagnetism and the weak force.
- **Proton-Driven Gravity**: Gravity, in this model, arises from the interactions among protons, particularly when they reach high densities. This gravitational effect isn't due to curved spacetime but emerges as a manifestation of the strong force in high-mass environments, particularly around massive celestial objects and black holes.
- **Electromagnetic and Weak Interactions as Proton-Influenced Phenomena**: Electromagnetic and weak nuclear interactions also derive from the proton's properties, modified by the proximity and distribution of electrons. This approach simplifies our understanding of forces, seeing them not as distinct entities but as expressions of a single underlying force.

This view of the proton as the source of all interactions creates a unified model of force that explains why particles and celestial bodies behave as they do, without requiring four separate fundamental forces.

Quarks as the Building Blocks of Dark Matter

Dark matter has remained one of the most elusive mysteries in physics. According to the Vetter Theory, the answer lies in quarks. Quarks are typically confined within protons and neutrons in ordinary matter, but in the vacuum of space, they exist freely and behave much like gas particles, filling the universe and providing the gravitational effects attributed to dark matter.

- **Free Quarks Filling the Universe**: Imagine quarks distributed throughout space similarly to how nitrogen fills the Earth's atmosphere. They exist everywhere, forming an invisible matrix of particles that we cannot directly observe because they lack electrons and therefore do not interact with light.
- **The Quark-Based Structure of Dark Matter**: These free quarks, undetectable by electromagnetic means, exert gravitational effects that help hold galaxies together, much like dark matter is observed to do. This explains the unseen mass in galaxies and clusters without needing exotic new particles.
- **Dynamic Quark Pressure and Stability**: The quark density in space is not static; it adjusts as matter is created and annihilated. When quark density changes, it creates variations in the "pressure" of these particles, which we experience as dark energy.

This perspective offers a tangible, particle-based explanation for dark matter and dark energy, weaving them into a unified framework that aligns with observable cosmic phenomena.

Matter Creation and Destruction: A Continuous Cycle

In conventional physics, matter is typically seen as stable and unchanging unless acted upon by specific reactions or processes. However, the Vetter Theory proposes that matter is created and destroyed constantly, forming a continuous cycle that shapes the universe's evolution. Black holes play a central role in this process.

- **The Role of Black Holes in Matter Cycling**: Black holes are often viewed as regions of endless density, but in the Vetter Theory, they function as proton collectors. As matter falls toward a black hole, it is accelerated to near-light speeds, causing the electrons and protons to separate. The protons are absorbed into the black hole, while the high-speed electrons are ejected outward into space.
- **Electron Ejection and Cosmic Recycling**: Ejected electrons are not simply discarded; they continue to move through space, free to recombine with quarks and form new particles. This constant cycling of matter—where electrons, protons, and quarks recombine and disperse—keeps the universe dynamic and evolving.
- **Matter Annihilation and the Quark Matrix**: The annihilation of matter, especially near black holes, affects the surrounding quark density, altering the "pressure" of quarks in space and influencing cosmic expansion.

This concept of matter cycling redefines our understanding of black holes and the behavior of matter across the cosmos. Rather than being static, the universe is a dynamic system with an ongoing exchange of particles and forces.

Dark Energy as the Pressure of Quarks in Space

Dark energy is known to drive the accelerated expansion of the universe, yet its nature remains mysterious. In the Vetter Theory, dark energy arises from changes in quark density, which affects the "pressure" exerted by these particles in the vacuum of space.

- **Quark Pressure as a Cosmic Force**: As matter is annihilated, the density and distribution of quarks change, influencing the "pressure" within the quark matrix. This pressure exerts an expansive force on the universe, driving it to expand.
- **Dynamic Influence on Expansion**: Unlike a static force, dark energy in this theory is dynamic, constantly fluctuating based on matter creation and destruction. This explains why the expansion of the universe isn't uniform and may vary over time.
- **A New Perspective on Cosmic Expansion**: By attributing cosmic expansion to quark pressure rather than mysterious energy fields, this theory provides a concrete, particle-based mechanism for dark energy, tying it directly to matter interactions and density fluctuations.

A Unified Model Rooted in Particle Interactions
The Vetter Theory offers a radically simplified view of the universe, where complex forces and mysterious phenomena are all variations of interactions between protons, quarks, and electrons. This model replaces the abstract, field-based theories of conventional physics with a direct, observable mechanism that extends seamlessly from particle scales to cosmic phenomena.

- **Bridging Quantum Mechanics and Cosmology**: By rooting all forces in the strong force and explaining dark matter and dark energy through quark behavior, the Vetter Theory creates a cohesive model that applies across scales, from subatomic particles to galaxies.
- **A Rejection of Complexity for Observability**: This theory challenges the need for complex mathematical constructs by offering an intuitive, particle-based view of the universe. By simplifying our understanding of forces and matter, the Vetter Theory seeks to make physics more accessible and grounded in observable reality.
- **An Open Invitation to Discovery**: As we explore these core principles in the coming chapters, the Vetter Theory will redefine our perspective on forces, matter, and the structure of the cosmos. It offers a fresh, cohesive vision of the universe that encourages us to rethink our place within it.
-

1.4 A New Perspective on Black Holes and Gravitational Lensing

Rethinking the Nature of Black Holes
In conventional astrophysics, black holes are viewed as regions of spacetime where gravity is so intense that nothing—not even light—can escape. These objects are often described as singularities, points of infinite density and zero volume, surrounded by an "event horizon" beyond which all information is lost. But the Vetter Theory offers a new, straightforward interpretation of black holes, removing the complexities of singularities and event horizons altogether.

In this model, a black hole is simply a dense, concentrated collection of protons. Rather than being an incomprehensible point of infinite density, a black hole is seen as a physical object—a proton collector—that plays a vital role in the cycling of matter throughout the universe. This approach demystifies black holes, making them tangible entities that fit seamlessly within the framework of particle interactions.

Black Holes as Proton Collectors

As matter approaches a black hole, it accelerates to near-light speeds due to the gravitational force resulting from the intense proton density. When matter reaches such extreme speeds, the particles begin to disintegrate. Electrons are stripped from their protons, leading to a split: protons are captured by the black hole, while electrons are ejected at incredibly high velocities.

- **Proton Accumulation**: The core of a black hole is formed by the accumulation of protons, which are collected and held due to their interactions with the intense gravitational field. This collection of protons creates a stable, dense structure rather than a singularity.
- **Electron Ejection**: Electrons, no longer bound to their protons, are expelled from the black hole, moving outward at near-light speeds. These high-speed electrons form a sort of "halo" around the black hole, continually streaming into space.

This redefinition of black holes as proton-rich bodies removes the need for singularities, event horizons, and spacetime warping. Instead, it frames black holes as physical objects with unique particle dynamics, grounded in observable and explainable interactions.

Gravitational Lensing Without Curved Spacetime

One of the most compelling phenomena observed around black holes is gravitational lensing, where light bends as it passes near these massive objects. Traditionally, this effect is explained by the curvature of spacetime around a black hole, causing light to follow a curved path. However, in the Vetter Theory, gravitational lensing is explained through a different mechanism—one involving high-speed electrons.

- **Electron Influence on Light**: In this framework, as electrons are ejected at near-light speeds from the vicinity of a black hole, they interact with passing photons. The intense speed and energy of these electrons create a field that influences the path of light, bending it in a way that appears similar to traditional gravitational lensing.
- **A Particle-Based Lensing Effect**: This particle interaction-based lensing doesn't rely on any bending of spacetime. Instead, it emerges from the electromagnetic and kinetic effects of these high-speed electrons on nearby photons. Light appears to bend around the black hole not because of curved space but because of the influence of electrons moving at extreme velocities.

This approach to lensing offers a fresh perspective, suggesting that observed light bending around massive objects is not due to spacetime distortion but is instead the result of interactions with particles moving near the speed of light. It redefines lensing as an effect grounded in particle physics, in line with the Vetter Theory's commitment to a simpler, more direct explanation of cosmic phenomena.

Implications of Black Hole Particle Dynamics

Redefining black holes and gravitational lensing in this way has profound implications for our understanding of the universe:

- **A Tangible Model for Black Holes**: Viewing black holes as dense proton accumulators simplifies their role in the cosmos, making them easier to study and understand. It eliminates the mysteries of singularities and event horizons, presenting black holes as dynamic objects that cycle matter and contribute to cosmic structure.
- **Predictable Lensing Patterns**: By tying lensing effects to high-speed electron behavior, this model can make specific predictions about how light will behave around black holes. If electron ejections influence light, then variations in electron density and speed should create distinct lensing patterns that can be observed and compared with traditional lensing predictions.
- **Alternative Explanation for Light Behavior Near Massive Bodies**: This particle-based approach suggests that other massive bodies, like neutron stars or dense galaxies, may also exhibit particle-driven lensing effects, challenging the notion that all light bending is due to spacetime curvature.

A Self-Contained System of Matter Cycling

In the Vetter Theory, black holes are not dead ends where matter is forever trapped. Instead, they are central players in a continual cycle of matter creation and destruction:

- **Electron Recycling**: Ejected electrons move outward from the black hole and continue to travel through space. Free to recombine with quarks in the vacuum, they contribute to the continuous regeneration of matter.
- **Black Holes as Catalysts for Matter Creation**: Rather than being ultimate endpoints, black holes are integral to the Vetter Theory's concept of a self-sustaining universe, constantly recycling particles and allowing matter to "refresh" and maintain cosmic equilibrium.

This perspective removes the concept of "information loss" associated with traditional black holes, replacing it with a model where matter and energy are continuously cycled back into the universe. Electrons, protons, and quarks are all participants in this process, each playing a role in maintaining the universe's dynamic balance.

Simplifying Our View of Black Holes and Lensing

By stripping away the abstract and complex ideas associated with black holes—singularities, spacetime curvature, and event horizons—the Vetter Theory offers a more accessible and grounded explanation for these enigmatic objects. Black holes are not mysterious portals or bottomless pits; they are physical entities, defined by particle interactions and characterized by the behavior of protons and electrons.

This model opens the door to a new way of observing and understanding black holes and gravitational lensing. If these effects result from particle behavior rather than spacetime curvature, our tools and methods for studying them could evolve, focusing more on particle dynamics and less on abstract mathematical constructs.

The Vetter Theory's perspective reimagines black holes as participants in the universe's natural cycles, not as isolated anomalies but as essential engines of matter recycling. It frames gravitational lensing as an interaction effect, reinforcing a consistent theme throughout the theory: that the universe operates on simple, direct principles rooted in particle interactions.

In the next chapter, we will explore how the vacuum of space itself—filled with free quarks behaving like an invisible gas—provides a basis for understanding dark matter. The cycling of matter around black holes, the influence of quarks in space, and the simplicity of the strong force's manifestations all contribute to a unified, comprehensible picture of the cosmos.

1.5 Bridging Micro-Scale Interactions with Cosmic Phenomena

A Universe Built on Particle Interactions

In the Vetter Theory, the universe is built on a set of simple, direct interactions among particles. This approach shifts away from viewing forces and cosmic phenomena as the result of abstract fields and spacetime geometry. Instead, the theory posits that the behavior of particles—specifically protons, quarks, and electrons—underlies everything from the forces governing matter to the structure of galaxies and the expansion of the universe itself.

At the core of this perspective is the idea that micro-scale interactions, those within and between particles, extend seamlessly to create the effects we observe on a cosmic scale. Rather than requiring a separate set of rules for the large and small, the Vetter Theory presents a unified framework where the same principles apply universally, providing an elegant, cohesive view of the universe.

The Strong Force as the Universal Force

In standard physics, the four fundamental forces each govern different types of interactions: gravity acts on mass, electromagnetism on charge, and the strong and weak forces within atomic nuclei. But in the Vetter Theory, the strong force is the only truly fundamental force, manifesting in different ways under different conditions to create the effects we recognize as gravity, electromagnetism, and the weak force.

- **Unified by the Proton**: All interactions trace back to the proton, where the strong force governs the behavior of quarks. At higher densities, this force radiates outward to create the effect of gravity, attracting other particles and giving shape to massive structures like stars and galaxies.
- **Electromagnetism and the Weak Force as Variants of the Strong Force**: The strong force modifies its effects based on electron distribution around protons, creating the electromagnetic and weak interactions as variations within the same fundamental force. This allows particles to interact in diverse ways while maintaining a consistent, singular force at the foundation.

This simplification removes the need for separate force-carrying particles or fields, replacing them with a single, adaptable force that directly governs all physical phenomena, from atomic bonding to the gravitational pull of galaxies.

Quarks as Dark Matter and the Structure of Space

Another key component of the Vetter Theory is the concept of quarks as dark matter. Dark matter has traditionally been considered an unknown form of mass that holds galaxies together, but in this framework, quarks are the essence of dark matter. These free quarks exist throughout space, filling it similarly to a gas and creating an invisible structure that helps define the universe's shape.

- **The Quark Matrix**: Imagine space as a vast field filled with quarks, moving and shifting like nitrogen in our atmosphere. These quarks are unseen because they lack electrons and do not interact with light, but they exert a gravitational influence on matter, binding galaxies and clusters together.
- **Gravitational Influence Without Mass Distortion**: Unlike relativity's dark matter model, which implies unseen mass warping spacetime, the Vetter Theory's quark matrix exerts gravitational influence directly. This interaction is entirely particle-based, eliminating the need for hidden mass or warped geometry.

This quark-based structure of space provides a clear and intuitive explanation for dark matter's effects, transforming an invisible "mystery mass" into a tangible and predictable component of the cosmos.

Dynamic Pressure and the Phenomenon of Dark Energy

Dark energy, the force driving the accelerated expansion of the universe, has puzzled physicists since its discovery. The Vetter Theory offers a novel explanation by tying dark energy to the "pressure" of quarks in the vacuum of space.

- **Matter Annihilation and Quark Density**: As matter is annihilated near black holes, quarks in the surrounding space adjust to fill the vacuum. This process changes the quark "pressure" in space, creating an expansive force that we experience as dark energy.

- **A Responsive, Self-Regulating Universe**: This model suggests that the universe's expansion isn't static but responds to matter creation and annihilation cycles. As quark pressure changes, it drives the cosmic expansion outward or inward, maintaining balance without requiring a mysterious energy field.

This particle-based approach reframes dark energy not as a force acting on spacetime, but as a result of particle density changes that influence cosmic expansion. It ties dark energy directly to matter interactions, uniting two of the universe's most enigmatic components—dark matter and dark energy—within a single, particle-driven framework.

How Micro-Scale Interactions Shape Galaxies and the Universe

In the Vetter Theory, the same basic principles governing particles also define galaxies, star systems, and large-scale cosmic structures. Here's how these micro-scale interactions translate to cosmic phenomena:

- **Galactic Structure**: Galaxies are bound and shaped by the quark matrix, which holds massive bodies in place without needing extra gravitational distortion. The proton-based gravitational attraction combines with quark "dark matter" density to stabilize galactic rotation and prevent stars from dispersing.
- **Black Hole Behavior and Electron Ejection**: The dynamics of black holes—the collection of protons and ejection of electrons—directly influence the surrounding quark density. High-speed electrons create lensing effects and modify quark density, affecting local gravitational fields and contributing to galaxy structure.
- **Cosmic Expansion**: Changes in quark density on a vast scale drive the universe's expansion. This expansion is not merely a relic of the Big Bang but an ongoing, responsive process affected by cycles of matter creation and destruction.

These dynamics allow us to observe a cosmos that is both vast and interconnected, with particles interacting across scales to maintain a cohesive structure. The universe's complexity is thus reduced to a few core interactions that govern everything from atomic behavior to the movement of galaxies.

A Simpler, Unified Framework for Physics

The Vetter Theory replaces the complexity of separate forces and fields with a single, adaptable framework. By grounding all interactions in particle behavior, it creates a model where the forces shaping particles also shape galaxies and cosmic structures, forming a continuous chain from the microscopic to the macroscopic.

- **Bridging Quantum Mechanics and Cosmology**: The proton-driven forces of this theory unite quantum mechanics with cosmic phenomena, seamlessly explaining phenomena across scales without the need for specialized physics at each level.
- **Observability Over Abstraction**: Unlike conventional theories that rely on abstract constructs like spacetime curvature and hidden mass, the Vetter Theory emphasizes observable phenomena. Every concept ties back to particles and their behaviors, creating a transparent, testable model that aligns with real-world observations.

- **Predictive Power**: This unified framework opens the door to new predictions. Since the quark matrix and proton-based gravity are predictable, they can generate testable hypotheses about galaxy rotation, cosmic expansion rates, and light behavior near massive objects. This predictive power challenges and builds upon current scientific understanding, suggesting future discoveries grounded in the Vetter Theory.

Toward a New Understanding of the Universe
As we continue through this book, we'll examine how each aspect of the Vetter Theory connects particle interactions with the structure and behavior of the cosmos. From the dense proton collectors we call black holes to the dynamic quark matrix that fills the universe, this framework replaces complexity with simplicity, creating a cohesive, observable, and testable model of physics.

By understanding the universe through particle interactions and proton-driven forces, the Vetter Theory offers a vision of the cosmos that is both elegant and grounded in simplicity. It invites us to rethink the relationship between the very small and the very large, offering a holistic perspective that redefines the forces, structures, and phenomena we observe in the universe.

In the next section, we'll dive into specific observational predictions that arise from the Vetter Theory, laying the groundwork for new experiments and discoveries that could transform our understanding of the cosmos.

1.6 What to Expect in This Book

Embarking on a Journey Through the Vetter Theory
Welcome to a new way of seeing the universe. Over the next chapters, you'll dive deeply into the Vetter Theory—a framework that challenges some of physics' most established ideas and proposes a more direct, intuitive understanding of the cosmos. This journey is about letting go of conventional assumptions and exploring a simpler, unified approach grounded in particle interactions and straightforward principles.

The Vetter Theory may ask you to suspend beliefs you thought were unshakable and embrace new ways of thinking. This isn't about complexity; it's about clarity. Whether you're an experienced scientist, a curious thinker, or someone with a passion for understanding how the universe works, this book will introduce you to a model that aims to connect every layer of the cosmos, from the smallest particles to the vastness of space.

Chapter Highlights: A Roadmap of Key Ideas
1. **Understanding the Proton's Central Role**: In the opening chapters, we'll explore the proton as the universe's fundamental building block and the origin of all forces. By understanding how the proton drives gravity, electromagnetism, and other interactions, we'll see how one particle can shape everything from atoms to galaxies.

2. **Redefining Dark Matter and Dark Energy**: Next, we'll look at quarks and their role as the essence of dark matter. You'll learn how the quark "matrix" fills the vacuum of space, holding galaxies together without mysterious or invisible mass. We'll also dive into dark energy, reinterpreted here as the "pressure" of quarks in space, dynamically influencing cosmic expansion.
3. **Black Holes as Matter Recyclers**: In later chapters, we'll shift our focus to black holes, where matter cycling reaches its peak. We'll explore how black holes function as dense proton collectors that eject high-speed electrons, influencing nearby light paths and creating effects traditionally attributed to gravitational lensing. This reframing positions black holes as active participants in the matter cycle rather than singularities or points of no return.
4. **Lensing and Light Behavior Without Curved Spacetime**: Another chapter will address gravitational lensing—showing how high-speed electrons affect the path of photons near massive objects, bending light in ways that mirror traditional lensing without requiring any curvature of spacetime. You'll see how particle interactions alone can explain these observations.
5. **A Self-Sustaining Universe**: We'll take a deeper look at the matter creation and destruction cycle in the Vetter Theory. This process maintains cosmic balance, ensuring that matter is continuously renewed and influencing the universe's structure on all scales. In this self-regulating universe, particles flow, recombine, and sustain cosmic harmony.
6. **Bringing Micro-Scale Interactions to Cosmic Structure**: Toward the end, we'll examine how this theory bridges the micro and macro scales, connecting particle behavior directly with the structure and expansion of the universe. By understanding the same principles across all scales, we'll achieve a cohesive view of physics that feels both intuitive and comprehensive.

An Invitation to Think Differently

As you read, you may feel compelled to question some of the core ideas that define modern physics. This book challenges the need for the complex, abstract concepts that are often difficult to visualize and verify. By focusing on the visible, observable, and testable, the Vetter Theory invites you to think about the universe in simple, intuitive terms.

This journey may change the way you think about particles, forces, and space itself. Rather than seeing the cosmos as a place of hidden dimensions and exotic forces, the Vetter Theory offers a perspective where everything is interconnected, rooted in the familiar, and governed by particle interactions we can trace and predict.

A Guide to the Theory's Practical Implications

Beyond simply explaining the universe's structure, the Vetter Theory has practical implications for science and technology. By reframing forces as particle-driven effects, this model could lead to new experimental techniques and breakthroughs in fields ranging from astrophysics to quantum mechanics. Whether it's a deeper understanding of particle behavior or new predictions about cosmic phenomena, the theory opens doors to practical discoveries that could reshape science.

Throughout this book, we'll not only dive into the theoretical aspects but also explore how these ideas could inspire experimental inquiries and novel approaches to understanding the universe. From cosmic lensing patterns to the behavior of particles near black holes, the Vetter Theory has the potential to drive real-world applications and inspire new research directions.

Keeping an Open Mind
The journey you're about to undertake isn't just about facts and figures; it's about perspective. The Vetter Theory challenges you to see the universe in a way that feels simpler, clearer, and more grounded in tangible principles. As we move through each chapter, consider setting aside preconceived ideas and exploring these concepts with an open mind. This book isn't just about challenging conventional physics; it's about presenting a vision of the universe that's accessible and rooted in common sense. The Vetter Theory may provide answers to questions we've long struggled to explain, but it also invites you, the reader, to participate in a new era of discovery.

An Invitation to Question and Explore
The Vetter Theory is more than a set of ideas; it's a challenge to every reader to think differently about the universe. In each chapter, you'll find ideas designed to provoke thought and inspire exploration. The beauty of science lies in its ability to evolve, and this book is an invitation to be part of that evolution.

As we move forward, keep in mind that this isn't just about the Vetter Theory; it's about opening doors to further understanding and discovery. By questioning, testing, and refining these ideas, you're helping shape the future of science. Whether you're a seasoned physicist or an interested thinker, this book welcomes you to explore, critique, and contribute to a deeper understanding of the universe.

Let's begin this journey with an open mind, a spirit of curiosity, and a willingness to reimagine the world around us.

Chapter 2

2.1 The Proton: The Fundamental Building Block of the Universe

The Proton as the Heart of the Universe

In conventional physics, the proton is often seen as a positively charged particle within atomic nuclei, essential for atomic structure but otherwise limited in scope. However, in the Vetter Theory, the proton takes on a much greater significance. It is not merely a particle but the foundational unit that generates all forces and structures in the universe. Through its unique interactions with quarks and electrons, the proton defines the behavior of matter on every scale—from the atomic to the cosmic.

At the heart of the proton lies a unique interaction between quarks and an electron. This relationship generates the strong force that binds quarks together, producing a stable structure shielded by the electron. This electron-quark interaction is essential, as it both creates and contains the strong force within the proton, allowing the proton to be the wellspring of all forces in the universe.

Quarks, the Electron, and the Generation of the Strong Force

Inside each proton are three quarks—two "up" quarks and one "down" quark. Traditional physics describes these quarks as being bound together by the strong force, mediated by gluons. However, the Vetter Theory introduces a new perspective: rather than gluons binding quarks, it is the interaction between the quarks and an electron that creates and contains the strong force.

- **Electron-Quark Interaction**: In the Vetter Theory, the strong force arises from the interaction between the electron and the three quarks within the proton. The electron's presence is crucial, as it insulates or "shields" the strong force, containing it within the proton.
- **A Self-Sustaining Force**: The electron acts as a stabilizing barrier around the quarks, allowing the strong force to self-sustain within the proton without breaking down. This unique insulation effect makes the proton one of the most stable particles in the universe, capable of maintaining its structure indefinitely under normal conditions.
- **The Insulating Electron**: The electron prevents the strong force from radiating outward unnecessarily, allowing the force to remain contained within the proton until specific conditions—such as proximity to other protons or high densities—allow it to manifest in different ways.

This electron-insulated structure of the proton allows the strong force to operate as a flexible and adaptable interaction, one that can extend beyond the proton under certain conditions to influence other particles and create the effects we observe as gravity, electromagnetism, and the weak force.

The Proton as the Source of All Fundamental Forces

In the Vetter Theory, the proton is not just a building block of matter but the origin of all fundamental forces. While conventional physics describes gravity, electromagnetism, and the weak and strong nuclear forces as distinct, the Vetter Theory posits that they are all manifestations of the strong force, produced and regulated by the proton.

- **The Strong Force as a Universal Force**: Within the proton, the strong force binds quarks together. However, under specific conditions, this same force can manifest differently to create gravitational, electromagnetic, or weak interactions.
- **Gravity as an Emergent Effect**: When protons accumulate in dense configurations, such as in massive stars or black holes, the strong force can radiate beyond the proton's electron insulation, creating the effect we experience as gravity. This emergence of gravity is directly tied to proton density rather than any curvature of spacetime.
- **Electromagnetic and Weak Interactions**: The electron's role in insulating the strong force also gives rise to electromagnetic effects, as the electron's movement around protons creates an electromagnetic field. Meanwhile, the weak interaction appears in particle interactions as another variant of the strong force, dependent on proximity and particle arrangements.

This simplified model unifies the fundamental forces, presenting them as diverse expressions of a single force originating from the proton. This view not only simplifies our understanding of forces but also aligns with observable phenomena in a cohesive, particle-based framework.

The Proton's Stability and Its Role in the Universe

One of the proton's defining characteristics is its remarkable stability, a feature that sets it apart from many other particles. Unlike neutrons, which decay when isolated, the proton maintains its structure over billions of years. This stability is not incidental; it is essential for forming the universe's complex structures.

- **Indefinite Stability**: The electron-quark interaction within the proton creates a stable, balanced structure that allows the proton to endure indefinitely. This durability is what enables protons to form the backbone of atoms, the building blocks of matter.
- **The Core of All Matter**: Every atom depends on the proton's stability to maintain its structure. Without this foundational stability, complex structures like molecules, planets, and stars could not exist.
- **Universal Consistency**: Protons are identical throughout the cosmos, whether they are within atoms on Earth or in the hearts of distant stars. This uniformity ensures that the laws governing matter are the same everywhere, providing a consistent framework for the universe to build upon.

By understanding the proton as more than a mere component of atoms, we see it as the universal building block, essential to the creation and stability of matter on all scales.

Proton-Driven Simplicity: A New Model of Forces

In traditional physics, each fundamental force requires its own theoretical framework and separate mediators, such as photons for electromagnetism or hypothetical gravitons for gravity. The Vetter Theory challenges this complexity by positioning the proton as the source of all forces, with the electron-insulated strong force manifesting differently based on particle interactions and conditions.

- **No Need for Separate Force Carriers**: In this model, there is no need for distinct particles to carry each force. The strong force alone, generated and insulated by the electron within the proton, adapts to produce gravitational, electromagnetic, and weak effects in a variety of contexts.
- **Self-Contained Interactions**: The proton's strong force, contained and regulated by the electron, eliminates the need for external fields or spacetime curvature to explain forces. This direct, particle-based approach allows for a simpler and more intuitive understanding of interactions.
- **A Unified, Predictive Model**: By rooting all interactions in the proton and its internal dynamics, the Vetter Theory provides a unified view that can explain a wide range of physical phenomena without relying on separate theories for each force. This cohesive framework enhances predictability and aligns well with observed reality.

A New Perspective on the Proton's Role in the Universe

In the Vetter Theory, the proton is much more than just a component of atoms; it is the key to understanding the universe itself. By generating and insulating the strong force, the proton becomes the source of all forces, shaping matter, energy, and cosmic structures. From the atomic level to the largest galactic scales, the proton's interactions define the behavior and organization of everything in existence.

As we move forward in this chapter, we'll explore how the proton's influence extends outward to create what we observe as gravity, electromagnetism, and the weak force, demonstrating how a single particle shapes the behavior of matter across the universe. This journey will reveal how, by focusing on the proton and its unique properties, we can unlock a unified understanding of forces, matter, and the cosmos itself.

This expansion captures the role of the electron-quark interaction in generating the strong force within the proton, aligning with your framework while setting the stage for the proton's central role. Let me know if this fits your vision or if there's any other detail you'd like to add!

2.2 The Strong Force as the Source of All Forces

The Hydrogen Atom: The Powerhouse of All Matter

In the Vetter Theory, the hydrogen atom isn't just the simplest element; it's the powerhouse behind all forces and the foundational unit of the universe. Uniquely stable and abundant, hydrogen consists of a single proton and electron whose interaction gives rise to the strong force. But more than that, hydrogen retains its essential identity even within heavier elements, serving as the enduring building block at the heart of all atomic structures.

While complex atoms incorporate additional protons and neutrons, the fundamental hydrogen structure—the interaction between a single proton and electron—persists. This unbreakable bond means that hydrogen, in essence, is embedded within the framework of every element, making it not only the origin but also the active force behind all matter, regardless of complexity.

The Strong Force Originates in the Proton, Manifests Through Hydrogen

Within the proton, three quarks are tightly bound, generating what we know as the strong force. In the Vetter Theory, this force does not arise from gluons but from the unique interaction between the proton's quarks and its insulating electron within the hydrogen atom. This electron-proton interaction allows the strong force to be contained, stored, and channeled in adaptable ways, with hydrogen's structure enabling it to manifest as the source of other forces.

- **Quark-Electron Interaction:** The interaction between the electron and the proton's quarks creates the strong force, with the electron playing an insulating role. This containment within hydrogen prevents the strong force from dissipating, giving it the stability needed to operate consistently across all matter.
- **The Role of Electron Shielding:** The electron shields the strong force, allowing it to remain focused within the hydrogen atom. This containment enables hydrogen to serve as a source of force, ready to be harnessed in different ways, depending on the context in which the atom resides.
- **A Flexible, Self-Contained Powerhouse:** Hydrogen's unique quark-electron configuration makes it a self-sustaining source of energy and stability. This adaptability allows the strong force within hydrogen to extend beyond the atom when conditions permit, giving rise to the varied forces observed in the universe.

The hydrogen atom's ability to maintain its core structure, even within more complex elements, ensures that the strong force and its adaptations persist in all atomic structures.

The Manifestation of Other Forces Within Hydrogen

While the strong force originates within the proton's quark-electron interaction, hydrogen is the true platform that allows this force to manifest as other interactions. The unique stability and adaptability of hydrogen enable it to produce gravitational, electromagnetic, and weak interactions, making it the powerhouse that sustains all observable forces.

- **Gravity as a Collective Manifestation of the Strong Force:** As hydrogen accumulates in dense environments, such as in stars or black holes, the strong force begins to radiate outward, creating the attraction we observe as gravity. This is not a warping of space but rather a direct result of the concentrated presence of hydrogen atoms, allowing gravity to emerge as an effect of cumulative proton density.
- **Electromagnetic Effects from Electron Behavior:** The electron within hydrogen doesn't merely insulate the strong force; it also generates electromagnetic interactions. The behavior of this electron, as it interacts with neighboring particles, gives rise to electromagnetic fields, demonstrating how hydrogen produces electromagnetism.
- **The Weak Force as a Transformation Within Hydrogen:** The weak force, responsible for certain types of particle transformations, appears in specific atomic interactions. In the Vetter Theory, the configuration within hydrogen is flexible enough to produce weak force effects, influencing particle decay and allowing hydrogen to adaptively drive transformations within matter.

This view of hydrogen as the source of all forces unifies gravity, electromagnetism, and the weak force, allowing them to be seen as adaptable expressions of the same underlying strong force.

Hydrogen's Persistent Identity Across All Elements

One of hydrogen's defining characteristics is that it retains its essential structure even within more complex elements. In heavier atoms, hydrogen doesn't lose its identity; rather, it serves as the core unit that builds upon itself, sustaining the structure and stability of all elements.

- **Hydrogen Embedded in All Matter:** Every atom, regardless of its complexity, contains hydrogen's fundamental structure—a proton-electron pair. This enduring identity makes hydrogen the backbone of all elements, ensuring that the same forces and interactions apply universally.
- **A Consistent Force Generator Across Scales:** Because hydrogen remains hydrogen within all elements, the strong force continues to operate consistently across different types of atoms. This gives rise to a cohesive structure within matter, where the same interactions apply universally.
- **Hydrogen as a Stabilizing Agent:** The stability and identity of hydrogen ensure that even in the densest and heaviest elements, the basic principles governing atomic interactions are preserved. This consistency provides the reliability needed for complex molecular structures and stable cosmic formations.

Hydrogen's persistent identity within all elements supports the Vetter Theory's premise that the universe's forces are rooted in a single, adaptable interaction rather than multiple independent forces.

Hydrogen as the Building Block of Cosmic Scale Interactions

In the Vetter Theory, hydrogen serves not only as a foundational unit within atoms but as the basis for all cosmic interactions. Its ubiquity and persistence across all elements allow hydrogen to bridge the micro and macro scales, connecting particle interactions to the formation of galaxies and stars.

- **Proton Density and Gravitational Influence:** Dense concentrations of hydrogen, such as in stars, produce gravitational effects through the accumulated strong force. This relationship between proton density and gravitational pull shapes large-scale structures, allowing hydrogen to be the binding force behind stars, planets, and galaxies.
- **Hydrogen's Role in Star Formation:** Hydrogen atoms are fundamental to the fusion processes that fuel stars. The interactions within hydrogen atoms produce the energy necessary for sustained fusion, making hydrogen both the fuel and the driving force of stellar processes.
- **Galactic Stability Through Hydrogen Distribution:** The abundance of hydrogen in galaxies provides a framework for galactic cohesion. Through the strong force, hydrogen atoms exert an attractive influence that binds stars and cosmic bodies, preventing them from drifting apart.

Hydrogen's persistence as the core building block across elements ensures that its forces scale seamlessly, from atomic structures to cosmic systems.

The Power of a Single Electron-Proton Bond in Hydrogen

At the heart of hydrogen is the simple, yet powerful bond between one proton and one electron. This bond doesn't just create a stable atom; it unleashes a controlled force that has the power to manifest as various interactions throughout the universe. In the Vetter Theory, this single electron-proton bond is the fundamental connection that drives all forces.

- **Electron Shielding and Force Modulation:** The electron in hydrogen shields and regulates the strong force, enabling the atom to store and manage energy with flexibility. This shielding gives hydrogen stability, allowing it to generate forces without losing its fundamental identity.
- **Adaptable Force Generation:** The electron-proton bond within hydrogen allows it to produce diverse forces, adapting the strong force as needed. This adaptability makes hydrogen the ultimate force generator, shaping interactions from atomic bonds to gravitational attraction.

By focusing on the hydrogen atom's unique bond, we see how this simple structure serves as the engine of energy, motion, and stability throughout the universe.

Hydrogen's Role in a Unified Force Model

In the Vetter Theory, hydrogen's single proton-electron structure is the central link that connects atomic interactions with cosmic phenomena. This simplicity enables a universal application of forces, creating a model where the same principles governing hydrogen apply throughout the universe.

- **From Atomic Bonds to Cosmic Scale Forces:** The hydrogen atom's ability to adapt its strong force interactions allows it to contribute to phenomena across scales, from molecular bonds to galaxy formation, achieving a universality that traditional models cannot.
- **A Model Rooted in Simplicity:** By placing hydrogen at the center, the Vetter Theory presents a streamlined model, where complexity is replaced by consistency. Hydrogen serves as the universal template, allowing us to view all forces and interactions as expressions of a single atomic structure.
- **Hydrogen as the Ultimate Connector:** The consistency of hydrogen across all matter provides a stable foundation that unifies physics on all scales. By understanding hydrogen as the engine behind all forces, we gain a comprehensive view of the universe's structure and operations.

Revolutionizing Our Understanding of Hydrogen and the Strong Force

In the Vetter Theory, hydrogen is not only the simplest atom but the universal force generator and the enduring foundation of all matter. By understanding how hydrogen remains constant within heavier elements, we reveal a universe that operates on consistent principles, where the same forces apply universally.

As we continue, we'll explore how hydrogen's role extends beyond the atomic scale, showing how its influence shapes the cosmos, from stars to galaxies. Through this lens, the hydrogen atom becomes not just a particle but the core element that drives the universe.

2.3 Gravity as a Manifestation of Proton Interactions

The Unique Origin of Gravity in the Vetter Theory

In conventional physics, gravity is explained as the result of mass curving spacetime, pulling objects together through a distortion of space and time. The Vetter Theory offers a completely different view. In this framework, gravity emerges from the collective behavior of protons, specifically through interactions within densely packed proton fields, as in stars or black holes. This is not due to any warping of spacetime but rather a direct extension of the strong force as proton density reaches a critical threshold.

- **Proton Density and Force Emanation**: When protons cluster in high-density configurations, the strong force that originates within each proton starts to extend beyond individual hydrogen atoms. This collective effect results in an attractive force on a larger scale, pulling nearby particles and objects toward the dense cluster of protons, creating what we experience as gravity.
- **Beyond the Proton Shell**: In ordinary conditions, the strong force remains contained within each proton, insulated by the electron field. However, in environments with intense proton clustering, such as within stellar cores or black holes, the strong force begins to operate beyond its typical boundary, creating a force that affects surrounding matter.

The Role of the Electron Field in Modulating Gravity

While the strong force emanates from protons, the electron field around them also plays a significant role. Electrons in surrounding atoms respond to proton density, and their behavior influences the intensity and reach of gravitational attraction.

- **Electron-Field Adjustment**: In high-density regions, electron fields around atoms adjust to the increased proton presence, effectively enhancing the gravitational influence of the proton cluster. This modulation is crucial for balancing gravitational strength and allowing for the unique gravity profiles observed in different types of celestial bodies.
- **Collective Electron-Field Effect**: The arrangement and movement of electrons around protons in densely packed atoms add a layer to the gravitational effect, reinforcing the attractive force. This interaction between proton density and electron configuration creates a unique gravity "signature" for every celestial body, contributing to its overall gravitational strength and reach.

Gravity as an Emergent Effect of the Strong Force

In this model, gravity doesn't arise as a fundamental force but as an emergent effect of the strong force, revealing itself only in specific high-density situations. This approach provides a more straightforward explanation for gravity, showing it as a byproduct of particle interactions rather than an independent force.

- **Emergence Through Density**: Gravity, in the Vetter Theory, is only evident when enough protons are in close proximity. As proton density increases, the cumulative strong force extends outward, creating a collective attraction that draws objects toward the center of the proton cluster.
- **Density Threshold for Gravity**: This model suggests that there is a threshold at which proton clustering becomes sufficient to produce a gravitational effect. Below this density, the strong force remains within the electron-shielded boundaries of individual hydrogen atoms, and no noticeable gravitational force emerges.

The Cosmic Implications of Proton-Based Gravity

This view of gravity changes how we understand the structure of stars, planets, and galaxies. In the Vetter Theory, cosmic bodies are bound together not by spacetime curvature but by the density-driven, emergent effect of the strong force from proton clusters. This creates a universe in which gravity is a dynamic, particle-driven force that varies depending on local proton density.

- **Gravity in Stars and Galaxies**: In stars, the dense hydrogen presence generates a strong gravitational effect, pulling matter inward and enabling fusion. Within galaxies, hydrogen's distribution creates gravity zones that keep stars in orbit, binding the galaxy together through a collective proton-generated force.
- **Black Holes as Extreme Proton Densities**: In black holes, proton density reaches such an extreme level that the gravitational pull becomes nearly insurmountable. Electrons are ejected, while protons cluster tightly, amplifying the gravitational effect to levels that prevent light from escaping.

Redefining Gravity: A Particle-Driven Approach
By viewing gravity as a result of proton interactions and density, the Vetter Theory offers a new model for understanding gravitational effects without the need for spacetime curvature.

- **Gravity as an Observable, Scalable Effect**: This approach makes gravity an observable outcome of proton clustering and electron-field modulation, allowing for precise predictions based on local proton densities and configurations.
- **Eliminating Gravitational Paradoxes**: By attributing gravity to particle density rather than an abstract field, this model removes the need for concepts like singularities and warping, providing a straightforward, particle-based explanation for gravitational attraction.
- **Consistency Across Scales**: The proton-based model of gravity applies universally, from atomic clusters to galaxies, offering a cohesive and adaptable framework for understanding gravitational phenomena.

In the Vetter Theory, gravity is not an intrinsic property of matter but an emergent effect of proton density, extending outward as an adaptation of the strong force under specific conditions. This understanding allows us to redefine gravity as a direct outcome of particle interactions, reshaping our perspective on cosmic structure and the fundamental forces governing the universe.

2.4 Electromagnetism and the Weak Force as Variants of the Strong Force

Rethinking Electromagnetism in the Vetter Theory

In traditional physics, electromagnetism is treated as a fundamental force, mediated by photons and governing charged particle interactions. However, in the Vetter Theory, electromagnetism is an adaptive expression of the strong force, arising from specific interactions between electrons and protons within the hydrogen atom and, by extension, within all matter.

The presence of electrons, orbiting or interacting with the nucleus, shapes the way the strong force manifests, creating electromagnetic effects without the need for a separate force carrier like photons. This reframing positions electromagnetism not as a standalone force but as a variant of the strong force, influenced by the electron's placement in relation to the proton.

- **Electron-Proton Configuration**: The distance and arrangement of electrons around protons dictate how the strong force operates, creating an electromagnetic field as electrons interact with each other and with protons. This electron-proton interaction allows the strong force to extend outward in the form of an electric field.
- **Charge as a Function of Electron Position**: In the Vetter Theory, what we perceive as electric charge is actually a manifestation of the electron's specific arrangement relative to the proton. As electrons shift, their changing positions modulate the strong force, producing observable electric and magnetic fields.
- **Magnetic Fields as Collective Electron Effects**: When electrons move in a coordinated manner, as in the case of current-carrying conductors, they create a magnetic field. This field is an extension of the strong force, adapted by electron motion rather than requiring a distinct electromagnetic force.

This interpretation eliminates the need for photons as force carriers, presenting electromagnetism as a dynamic form of the strong force shaped by the electron's influence within the hydrogen atom.

Electromagnetism in Complex Atoms and Materials

In more complex atomic structures, where multiple protons and electrons interact, the influence of the strong force extends to create complex electromagnetic behaviors. This variation in electromagnetic effects allows for a range of properties in different elements and materials.

- **Electron Field Interactions in Larger Atoms**: In atoms with multiple protons and electrons, the arrangement and motion of electrons in their fields amplify or reduce electromagnetic effects. This results in elements with varying degrees of conductivity, magnetism, and charge interactions.
- **Material Properties as Strong Force Variants**: Metals, insulators, and other materials exhibit unique properties because of how their atomic structures interact with the strong force. Electrons in metals, for example, are free to move and create fields, while in insulators, electron positioning restricts field generation, resulting in different electromagnetic behaviors.

- **Coordinated Fields in Molecules**: When atoms bond, the arrangement of electrons across molecules creates a collective electromagnetic field that drives molecular interactions. The Vetter Theory presents these molecular bonds as manifestations of the strong force, regulated by electron configuration rather than separate electromagnetic interactions.

This explanation unifies material properties and chemical bonds under a single framework, with the strong force providing the basis for all electromagnetic effects observed in various states of matter.

The Weak Force as Another Manifestation of the Strong Force

The weak nuclear force, traditionally associated with particle decay and radioactive processes, is also a variation of the strong force in the Vetter Theory. Within this framework, the weak force appears as a transformation mechanism within certain atomic and subatomic conditions, allowing particles to change form or decay.

- **Particle Decay Through Proton-Electron Interactions**: When protons and electrons in certain configurations experience instability, the strong force facilitates particle transformations. This transformation process manifests as the weak force, enabling phenomena like beta decay and other nuclear reactions.
- **A Controlled Adaptation of the Strong Force**: The weak force does not require a separate mediating particle, such as a W or Z boson. Instead, it arises from specific interactions between electrons and protons under high-energy or unstable conditions, allowing particles to rearrange or convert according to the strong force's dynamics.
- **Weak Force Effects in Radioactive Elements**: Radioactive decay in elements such as uranium or plutonium results from an imbalance in proton-electron configurations. When the strong force within these configurations reaches a threshold, it induces transformations that emit particles and release energy.

This interpretation of the weak force removes the need for additional force carriers, positioning it as another adaptive effect of the strong force, activated under specific nuclear conditions.

Unifying Electromagnetic and Weak Interactions Under a Single Force

In the Vetter Theory, both electromagnetism and the weak force are not distinct forces but different expressions of the strong force. The adaptability of the strong force—based on electron positioning and nuclear configuration—allows it to produce a range of effects without requiring separate theories or mediators.

- **Electromagnetism as Force Adapted by Electron Motion**: Electromagnetic fields arise from the positioning and movement of electrons, which modulate the strong force around the proton. This adaptability allows the strong force to manifest as electric or magnetic fields, depending on how electrons are arranged and interact.

- **The Weak Force as a Threshold Effect**: The weak force emerges when proton-electron interactions within certain nuclear configurations reach an energy threshold, enabling particle decay and transformation. This adaptation of the strong force facilitates radioactive decay and particle transformations without needing a unique weak force mediator.
- **A Cohesive Framework of Interactions**: By explaining both electromagnetism and the weak force as variants of the strong force, the Vetter Theory simplifies our understanding of atomic and subatomic interactions, unifying traditionally separate phenomena under a single, versatile force.

This unified approach presents a cleaner, more integrated model of forces, showing that what we observe as distinct interactions are simply adaptations of the same foundational force.

Implications for Atomic and Molecular Behavior

By positioning electromagnetism and the weak force as adaptive forms of the strong force, the Vetter Theory redefines how we understand atomic and molecular interactions. This perspective has wide-ranging implications for fields from chemistry to materials science.

- **Chemical Bonds as Strong Force Interactions**: In this model, molecular bonds are explained as configurations of the strong force, stabilized by electron placement. Covalent and ionic bonds are seen as expressions of the strong force adapted by electron-sharing or transfer, creating stable molecular structures without needing separate electromagnetic interactions.
- **Material Conductivity and Magnetic Properties**: Metals, insulators, and magnets exhibit distinct properties because of their unique electron configurations and the way these configurations allow the strong force to manifest as electromagnetic effects. This interpretation simplifies material science, linking conductivity and magnetism directly to proton-electron configurations.
- **Predictability Across Elements**: With electromagnetism and the weak force unified under the strong force, the Vetter Theory offers a predictable framework for understanding elemental behavior. Elements display specific properties based on electron arrangements around protons, allowing for a more consistent understanding of atomic and molecular behavior.

This model provides a comprehensive approach to understanding both chemical and physical properties, positioning them as effects of a single adaptable force rather than multiple separate interactions.

Electromagnetism and the Weak Force: Extensions of the Strong Force

In the Vetter Theory, the diversity of interactions in the universe is a result of the strong force's adaptability. By configuring electron placement and proton clustering, the strong force can manifest as electromagnetism, the weak force, and even gravity. This redefinition of forces offers a straightforward, particle-based explanation for complex phenomena, eliminating the need for separate force theories and allowing a unified view of matter.

In the following sections, we'll continue to explore how this unified approach influences the behavior of particles, atoms, and cosmic bodies, demonstrating how the strong force drives every interaction in the universe.

2.5 Proton Interactions Across Scales: From Atoms to Galaxies

From Atomic Structure to Universal Organization
In the Vetter Theory, the proton is the foundational unit whose interactions, modulated by the strong force and influenced by electron arrangements, extend seamlessly from atomic-scale forces to the gravitational interactions that shape galaxies. By understanding protons as the carriers of the strong force, we see that the same principles that govern atomic stability also govern the structures of stars and galaxies. This continuity across scales offers a unified view of the universe, with the proton's behavior at its heart.

Atomic Structure and Stability Through Proton Interactions
At the atomic level, the proton's interactions within hydrogen atoms establish a stable foundation for all matter. This stability is key to atomic cohesion, providing the reliability needed to create complex atomic and molecular structures.

- **Hydrogen as the Atomic Foundation**: As we've discussed, hydrogen's unique electron-proton interaction allows it to be the core component of all matter, with its single proton acting as a stable source of the strong force. This structure is preserved even in heavier elements, where additional neutrons space out protons without altering the proton's fundamental role.
- **Molecular Bonding and Proton-Electron Fields**: The arrangement of protons and electrons within atoms dictates how molecules form and interact. Covalent, ionic, and metallic bonds can all be seen as expressions of the strong force within and between atoms, stabilized by electron-proton fields.
- **Material Properties and Stability**: The unique arrangement of protons and electrons in each atom influences its electromagnetic properties, creating the diversity of materials and elements observed in nature. Conductivity, magnetism, and bonding types are all dictated by the way the strong force operates across atomic structures.

This atomic-level behavior provides the basis for all matter, establishing the foundation upon which larger cosmic structures are built.

Extending Proton-Based Forces to Stellar and Planetary Structures
The same forces that stabilize atomic structures also play a role in organizing matter at stellar and planetary scales. In stars, the proton density becomes especially critical, allowing the strong force to manifest as gravity and drive the formation of these massive bodies.

- **Star Formation Through Proton Density**: In stellar formation, hydrogen accumulates in massive quantities, increasing proton density to a critical level where the strong force extends outward as gravity. This gravitational force pulls matter inward, creating a self-sustaining structure that ignites nuclear fusion and powers the star.
- **Nuclear Fusion as a Proton-Driven Process**: The fusion process itself is a result of proton interactions. Under intense pressure and temperature, protons overcome their electron-shielded boundaries, allowing them to fuse and release energy. This process sustains stars and contributes to the creation of heavier elements.
- **Planetary Cohesion and Stability**: Around stars, planets form from dense clouds of hydrogen and other elements. The strong force within these atomic structures allows planets to maintain their cohesion, with gravity emerging from the cumulative effect of proton density to hold planets and their atmospheres together.

This scaling of the strong force to larger bodies creates stability in stars and planets, allowing them to persist over billions of years as stable, gravitationally bound structures.

Galactic Structure and Proton Density

At the galactic scale, hydrogen's ubiquity plays a fundamental role in organizing stars and planetary systems. The combined effect of proton-based interactions at this scale creates gravity wells that bind galaxies together and define their shape and movement.

- **Galaxies Bound by Proton-Driven Gravity**: The dense hydrogen presence within galaxies creates large-scale gravitational zones that bind stars into coherent structures. Each galaxy's gravitational pull results from the cumulative proton density within its vast clouds of hydrogen, acting as a unifying force that keeps stars in orbit around galactic centers.
- **Hydrogen and Dark Matter**: In the Vetter Theory, free quarks in space serve as dark matter, exerting gravitational effects without being visible. These quarks, which fill the space between stars and galaxies, interact with hydrogen's proton density to create the additional gravitational "glue" that holds galaxies together.
- **The Galactic Core as a Dense Proton Collection**: At the center of many galaxies lies a dense cluster of hydrogen and possibly even black holes, where proton density reaches extreme levels. This high proton concentration produces the gravitational core around which galaxies orbit, preventing stars from dispersing outward.

This model of galactic structure unifies the gravitational effects observed in stars, planets, and galaxies, showing how proton density creates stable cosmic structures through the strong force.

The Cosmic Scale and Black Hole Interactions

In the densest regions of the universe, where black holes form, proton density reaches extreme levels, creating gravitational effects that are both intense and far-reaching. In the Vetter Theory, black holes are dense proton accumulators that produce powerful gravitational fields as a direct result of proton interactions.

- **Black Holes as Proton Accumulators**: Rather than being points of infinite density, black holes are dense clusters of protons, with each proton contributing to the cumulative gravitational pull. In these extreme environments, electrons are ejected while protons cluster, amplifying the gravitational force to the point where even light cannot escape.
- **Electron Ejection and Light Bending**: As electrons are ejected from black holes, their high-speed motion influences nearby photons, bending light and creating effects similar to gravitational lensing. This effect, however, is not due to spacetime curvature but to direct particle interactions between high-speed electrons and photons.
- **Matter Cycling and Quark Pressure**: Black holes play a key role in the cycling of matter, collecting protons and influencing quark density in surrounding space. This cycling affects the "pressure" of quarks in the quark matrix, influencing cosmic expansion and the formation of dark energy.

By redefining black holes as proton-dense regions rather than singularities, the Vetter Theory provides a particle-based explanation for their gravitational behavior and influence on surrounding space.

A Unified Picture of Forces and Structures Across Scales

In the Vetter Theory, every scale of structure, from atoms to galaxies, is governed by the same underlying principles of proton-based forces. This unifying view simplifies our understanding of the universe, showing that the same particle interactions apply universally.

- **The Consistency of Proton-Based Gravity**: Whether at the atomic or galactic level, gravity emerges as a result of proton density, with the same principles governing all scales of matter. This consistency creates a predictable, unified framework for understanding gravity as a particle-driven force.
- **Electromagnetic and Weak Effects Across Structures**: The same principles that govern electromagnetic and weak interactions at the atomic level also apply to larger scales, influencing molecular bonding, material properties, and magnetic fields within planets and stars.
- **Galactic and Cosmic Balance Through Proton Density**: By positioning proton density as the basis of gravity, the Vetter Theory offers a balanced, particle-driven model that explains why galaxies, stars, and planetary systems are stable, cohesive structures.

Proton Interactions as the Foundation of Cosmic Order

The Vetter Theory unifies the behavior of forces across all scales by showing how proton interactions, modulated by the strong force, shape every structure in the universe. From the smallest atoms to the largest galaxies, the same principles apply, providing a cohesive view that links the micro and macro scales in a single, particle-driven framework.

In the following section, we will explore how this unified model allows the universe to self-regulate, maintaining stability through balanced proton interactions, quark density, and matter cycling. This self-regulating aspect of the universe highlights the role of proton-based forces as the underlying drivers of cosmic evolution.

2.6 The Proton's Role in a Self-Regulating Universe

A Dynamic Yet Balanced Cosmos

In the Vetter Theory, the universe isn't static; it's in a constant state of renewal, powered by the behavior of protons and quarks. Unlike conventional models that rely on unchanging structures or external forces, this framework presents a universe that self-regulates through the interactions of particles and forces. The proton, as the foundational particle, plays a critical role in maintaining this balance, driving both stability and dynamic evolution.

By continuously cycling matter, balancing proton density, and regulating quark pressure, the universe sustains itself without the need for external intervention. This model of a self-regulating cosmos suggests that every particle interaction contributes to a larger cosmic equilibrium, where matter and energy are recycled and transformed, allowing the universe to expand and evolve.

Protons as Catalysts in Matter Creation and Annihilation

In the Vetter Theory, matter is not fixed but is constantly created and destroyed in a continuous cycle. Black holes and other high-density regions are essential to this process, where protons play a critical role in both gathering matter and facilitating its transformation.

- **Black Holes as Proton Collectors**: Black holes act as dense proton accumulators. As matter accelerates toward black holes, electrons are stripped away, leaving protons to be absorbed into the core. This process creates high-density proton clusters that amplify gravitational effects while cycling matter through the universe.
- **Electron Ejection and Matter Renewal**: As electrons are ejected from black holes, they are free to recombine with quarks in space, creating new hydrogen atoms and contributing to a cycle of matter creation. This process continually refreshes cosmic matter, maintaining a balance of particles in the universe.
- **Matter Transformation and Reformation**: Protons collected in black holes are not destroyed but play a role in the recycling of matter, participating in the constant flow of particles throughout the universe. This matter transformation prevents the universe from becoming stagnant, enabling ongoing evolution and expansion.

This matter cycling, driven by protons, allows the universe to continually refresh itself, creating a dynamic equilibrium that supports cosmic growth and transformation.

Balancing Quark Density and Dark Energy

In the Vetter Theory, quarks exist freely in the vacuum of space, forming what we perceive as dark matter. These quarks exert pressure on the universe, influencing its expansion. As matter is created and annihilated, the density of quarks adjusts accordingly, affecting the "quark pressure" that contributes to cosmic balance.

- **Quark Density and Dark Matter**: Free quarks in space create a "quark matrix," filling the universe in a manner similar to how gases fill a container. This matrix acts as dark matter, exerting gravitational influence without visible mass, providing stability to galaxies and clusters.
- **Dark Energy as Quark Pressure**: When matter is annihilated, particularly near black holes, quark density adjusts, creating a pressure that contributes to the force we perceive as dark energy. This "quark pressure" drives cosmic expansion, adjusting as matter density fluctuates and allowing the universe to expand at a regulated pace.
- **A Self-Balancing Quark Matrix**: The quark density in space automatically adjusts in response to the creation and annihilation of matter, creating a self-balancing system that regulates dark energy. This balance ensures that cosmic expansion remains stable over time, driven by particle interactions rather than external forces.

This dynamic interaction between protons and quarks enables a universe where both dark matter and dark energy are particle-driven phenomena, contributing to a naturally regulated cosmic balance.

Proton-Driven Gravity and Cosmic Cohesion

The Vetter Theory's model of gravity, emerging from proton density, creates stability across cosmic structures without relying on spacetime curvature. Proton-based gravity binds stars, planets, and galaxies, creating a cohesive universe where proton density directly influences gravitational effects.

- **Gravitational Balance in Stars and Galaxies**: The cumulative proton density within stars and galaxies generates gravity, binding matter and creating stable cosmic structures. This density-driven gravity provides the cohesion necessary to maintain the integrity of galaxies and planetary systems.
- **Galaxy Formation and Proton Clustering**: As galaxies form, clusters of hydrogen (and thus protons) create gravity wells that bind stars together. This gravitational effect prevents stars from drifting apart, ensuring galactic stability while allowing for movement and rotation within the galaxy.

- **Cosmic Scale Stability**: Proton-driven gravity maintains balance on a universal scale, holding together vast structures and ensuring stability across space. This particle-based gravity provides a cohesive structure for the universe without needing to invoke spacetime distortion.

By grounding gravity in proton interactions, the Vetter Theory provides a stable framework for cosmic cohesion, allowing for both structure and movement on a universal scale.

A Universe in Equilibrium: Proton Interactions and the Self-Regulation of Forces

In the Vetter Theory, the universe doesn't rely on independent forces; instead, it uses proton interactions to create a self-regulating system where all forces are extensions of the strong force. This adaptability ensures that the universe can balance itself, creating stability through natural interactions.

- **Electromagnetism and the Weak Force as Adaptations**: The strong force adapts based on the arrangement of protons and electrons, manifesting as electromagnetism, gravity, and the weak force. These adaptations ensure that forces operate seamlessly across scales, from atomic to galactic.
- **Balanced Expansion Through Quark Pressure**: Cosmic expansion is driven by the pressure of quarks in space, influenced by proton density and matter cycling. This quark-based dark energy regulates the pace of expansion, ensuring that the universe remains balanced even as it grows.
- **Self-Regulation Without External Forces**: By rooting all interactions in proton dynamics, the Vetter Theory offers a model where the universe maintains equilibrium without needing external interventions or adjustments. Every cosmic event, from star formation to black hole behavior, contributes to a naturally balanced system.

This framework creates a universe that is both stable and flexible, capable of expanding, evolving, and adjusting as needed to maintain cosmic order.

Protons as the Key to Cosmic Sustainability

The Vetter Theory's model of a self-regulating universe suggests that protons are not only the foundation of matter but the catalysts of cosmic sustainability. By continuously cycling matter, balancing quark density, and adapting forces, the universe maintains itself as a cohesive, evolving system.

- **A Universe That Renews Itself**: Through proton-driven matter cycling and quark-based expansion, the universe remains in a state of constant renewal, preventing decay and ensuring that matter and energy are perpetually available.
- **Quark Dynamics and Long-Term Stability**: The adaptable quark density, responding to proton interactions, provides the cosmic "glue" that prevents galaxies from dispersing while supporting expansion. This quark-driven balance creates a stable, yet dynamic, structure.

- **Simplicity and Cohesion Across Scales**: By focusing on the proton as the source of all interactions, the Vetter Theory creates a model that is both simple and cohesive, explaining cosmic phenomena with a single particle-centered framework.

A Self-Regulating Universe Powered by Proton Dynamics

In the Vetter Theory, the universe doesn't rely on mysterious or external forces; it maintains itself through natural proton interactions and quark dynamics. This self-regulating model creates a cosmos that is not only sustainable but continually evolving, balanced through particle-driven interactions that extend across all scales.

As we continue, we will explore how this balanced framework leads to observable patterns and phenomena, from the behavior of light near black holes to the cosmic expansion rate. The Vetter Theory's particle-driven model of cosmic self-regulation provides a new way of understanding the universe as a system that is inherently capable of maintaining balance and order.

2.7 Summary: The Proton as the Universe's Master Architect

Recapping the Proton's Central Role in the Cosmos

Throughout Chapter 2, we've explored the profound significance of the proton within the Vetter Theory. More than just a subatomic particle, the proton is revealed as the architect behind every force, interaction, and structure in the universe. Together with its electron counterpart in the hydrogen atom, the proton drives a cohesive framework that unifies atomic stability, material diversity, and cosmic scale phenomena.

This new perspective positions the proton as the source of all fundamental forces, with the strong force adapting to produce gravitational, electromagnetic, and weak interactions. Through its behavior and interactions, the proton not only builds the atomic and molecular structures we observe but also provides the mechanisms behind stellar formation, galactic cohesion, and the universe's expansion.

The Strong Force as the Root of All Forces

In the Vetter Theory, the strong force within the proton is responsible for all other forces. Rather than viewing gravity, electromagnetism, and the weak force as independent, we've seen that they are different manifestations of the strong force, influenced by electron positioning and proton density.

- **Gravity as an Emergent Effect**: Gravity emerges in high-density environments, such as stars and black holes, as the cumulative effect of proton clustering. This interpretation provides a particle-driven explanation for gravity, rooted in proton density rather than spacetime curvature.
- **Electromagnetism Through Electron Interactions**: The arrangement of electrons around protons modulates the strong force, creating electric and magnetic fields. This redefinition of electromagnetism as an effect of proton-electron interactions eliminates the need for separate force carriers like photons.

- **The Weak Force as a Transformation Threshold**: Particle decay and transformation processes arise from instability within certain proton-electron configurations, allowing for the adaptive behavior of the strong force within atoms and molecules.

This unified force model simplifies our understanding of interactions, positioning the proton as the core driver of all observed forces.

Hydrogen as the Foundation of All Matter

The hydrogen atom, with its single proton-electron configuration, is the true powerhouse in the Vetter Theory. Hydrogen retains its essential structure even within heavier elements, making it the universal building block for all matter. This simplicity provides the foundation for everything from atomic stability to cosmic organization.

- **Universal Consistency Across Elements**: Hydrogen retains its identity even in complex elements, where additional neutrons serve to space protons apart while preserving the proton-driven strong force. This continuity ensures that the principles governing matter are uniform across scales.
- **Building Complex Structures**: The arrangement of protons and electrons within atoms influences material properties, from conductivity to magnetism, unifying chemical and physical behaviors under a single, proton-centered framework.
- **The Role of Hydrogen in Cosmic Balance**: Hydrogen's ubiquity ensures that proton-driven forces apply universally, allowing the same basic principles to govern everything from atomic cohesion to galactic stability.

By understanding hydrogen as the basis for all interactions, we reveal how its proton-driven forces create a stable yet adaptable universe.

Protons and Quarks as the Source of Dark Matter and Dark Energy

The Vetter Theory also reinterprets dark matter and dark energy through the lens of proton and quark behavior. In this model, free quarks in space form a matrix that creates the gravitational effects attributed to dark matter, while quark density fluctuations drive cosmic expansion as dark energy.

- **Quark Density as Dark Matter**: Free quarks in the vacuum of space fill the universe in a quark matrix, providing the gravitational "glue" that holds galaxies and clusters together without being visible.
- **Quark Pressure as Dark Energy**: Quark density adjusts as matter is created and annihilated, producing a pressure that drives cosmic expansion. This "quark pressure" provides a dynamic, particle-based explanation for dark energy, balancing cosmic forces as the universe expands.
- **Self-Balancing Matter and Energy**: This framework creates a self-regulating universe where proton density, quark pressure, and matter cycling maintain stability across cosmic structures.

These reinterpretations of dark matter and dark energy reinforce the proton's central role in maintaining a balanced, cohesive universe.

The Proton as a Catalyst for a Self-Regulating Universe

The proton, through its interactions and effects on quarks, drives a continuous cycle of matter creation and transformation. This dynamic behavior ensures that the universe is constantly renewed, maintaining equilibrium without external intervention.

- **Matter Cycling and Cosmic Renewal**: Black holes and other high-density regions play a role in matter cycling, gathering protons and ejecting electrons to contribute to a cycle of creation and annihilation. This process sustains a dynamic universe where matter is continually refreshed.
- **Self-Regulated Expansion and Stability**: Through quark density and proton-driven gravity, the universe maintains cosmic balance, allowing for stability and growth simultaneously. This self-regulation eliminates the need for exotic forces, providing a natural equilibrium.
- **A Dynamic but Balanced Universe**: Every particle interaction contributes to a larger cosmic order, where the balance of forces emerges naturally, creating a universe that is both stable and adaptable.

This view presents the universe as a self-sustaining system where protons and quarks collaborate to maintain balance across all scales.

The Proton as the Foundation of a Unified Universe

In the Vetter Theory, the proton's role transcends atomic structure, extending to every layer of the universe's architecture. From enabling atomic cohesion to driving cosmic expansion, the proton is the linchpin that unites forces, organizes matter, and sustains balance across all scales.

As we move forward, we will build on this foundation to explore the observable predictions and practical implications of the Vetter Theory, examining how this proton-centered model can enhance our understanding of the universe and guide future discoveries.

Chapter 3

3.1 Gravitational Lensing: Light Bending Without Curved Spacetime

Rethinking Gravitational Lensing in the Vetter Theory

In traditional physics, gravitational lensing—the bending of light around massive objects like galaxies and black holes—is understood as an effect of curved spacetime, where mass distorts space and forces light to follow a curved path. The Vetter Theory, however, offers a particle-based perspective on lensing, rooted in the high-speed interactions of electrons and photons near dense proton clusters. This approach eliminates the need for spacetime curvature, suggesting instead that light bending is an observable outcome of particle dynamics.

In this model, the intense proton density within massive objects like black holes results in the ejection of electrons at near-light speeds. As these high-speed electrons interact with photons, they create a bending effect on light's trajectory. This particle-based interaction provides a straightforward explanation for gravitational lensing, grounded in observable particle dynamics rather than abstract field distortions.

High-Speed Electron Ejections and Photon Path Distortion

The Vetter Theory's approach to gravitational lensing centers on the behavior of electrons ejected from dense proton clusters. Near objects like black holes, proton density reaches extreme levels, producing a gravitational pull so strong that it accelerates nearby matter, particularly electrons, to near-light speeds. These ejected electrons influence the path of passing photons through particle interactions, creating a bending effect on light.

- **Electron-Photon Interactions**: As high-speed electrons are ejected, they generate a field that alters the path of photons moving nearby. When photons pass through this electron-rich field, they experience a change in direction due to direct particle interactions, similar to light refraction.
- **Creating a Lensing Effect**: This particle-based bending of light mimics the effect traditionally attributed to curved spacetime. Instead of light following a geodesic around a massive object, photons are influenced by the energetic presence of electrons moving at extreme velocities.
- **Localized Particle Fields Rather Than Universal Curvature**: Unlike gravitational lensing based on spacetime curvature, this model suggests that lensing is a local effect, determined by the density and velocity of electrons around massive objects. This approach simplifies our understanding of lensing, grounding it in particle dynamics.

This mechanism for lensing repositions the phenomenon as a product of electron-photon interactions, providing an alternative to the idea of a universally curved spacetime.

Implications for Black Hole Observations

In regions near black holes, where gravitational effects are most intense, this electron-driven model of lensing offers distinct observational predictions. The behavior of light near black holes in the Vetter Theory may differ in subtle ways from traditional expectations based on spacetime curvature, providing unique observational opportunities.

- **Light Behavior in Electron-Dense Fields**: The Vetter Theory suggests that in high-electron fields, photons may experience variable degrees of bending depending on electron velocity and density. This could lead to differences in lensing effects near black holes, particularly in areas of high electron ejection.
- **Observable Differences from Curvature-Based Lensing**: Unlike the smooth curvature predicted by general relativity, particle-driven lensing might produce variations or asymmetries in the observed bending of light around black holes, potentially detectable with advanced telescopes.

- **Photon Redirection Patterns**: As photons are redirected by electron fields, there may be patterns or specific trajectories observable in light paths around black holes that diverge from conventional lensing models. This could provide a means of testing the Vetter Theory's predictions against observed lensing phenomena.

These implications open the door for further exploration, allowing for experimental comparisons between particle-driven lensing and traditional curvature-based models.

Experimental Predictions for Lensing in the Vetter Theory

The particle-based model of gravitational lensing in the Vetter Theory leads to several testable predictions, particularly regarding the behavior of light around black holes and other dense objects. These predictions suggest observable differences that could provide insight into the validity of the particle-based approach.

- **Asymmetries in Lensing Patterns**: If lensing is driven by particle dynamics, then variations in electron density and velocity around massive objects should result in asymmetrical or non-uniform lensing patterns, particularly in complex or fluctuating fields around black holes.
- **Intensity-Based Variations in Lensing**: The degree of bending should correlate directly with electron density and energy, producing observable intensity variations in lensing effects around different types of massive objects, such as neutron stars, black holes, or dense galactic cores.
- **Delayed Photon Arrival Times**: Particle interactions near high-density fields may cause slight delays in photon travel times, especially in regions where photons interact extensively with high-speed electrons. This could produce detectable arrival time differences in photons that pass near dense gravitational fields.

These predictions provide a path for testing the Vetter Theory's model of gravitational lensing, offering a way to compare particle-driven lensing with curvature-based interpretations in observational astronomy.

Redefining Gravitational Lensing: Light as a Result of Particle Interactions

The Vetter Theory's interpretation of gravitational lensing offers a particle-driven alternative to the concept of spacetime curvature. By positioning lensing as an effect of high-speed electron-photon interactions, this approach redefines light bending as an observable, local phenomenon influenced by particle behavior rather than field distortions.

- **Lensing as an Observable Phenomenon**: Unlike the abstract concept of curved spacetime, particle-driven lensing can be directly tied to observable variables like electron velocity and density. This makes lensing a testable outcome of particle dynamics, providing a straightforward and tangible interpretation.

- **A Model Grounded in Particles Rather Than Fields**: The Vetter Theory's approach to lensing eliminates the need for hypothetical field distortions, presenting light bending as an effect directly resulting from particle behavior. This localized model creates a consistent, particle-centered view of lensing that aligns with the theory's broader framework.
- **Potential Implications for Astrophysics**: If confirmed, this approach could reshape our understanding of black holes, neutron stars, and other dense celestial objects. It would provide a new lens through which to view the universe, grounded in particle interactions rather than abstract curvatures.

Toward a New Understanding of Light Behavior in the Cosmos
Gravitational lensing in the Vetter Theory transforms our understanding of light behavior around massive objects. By reinterpreting light bending as a particle interaction effect, this model offers a cohesive, testable approach to lensing that aligns with observable phenomena. This view opens new avenues for research and exploration, challenging us to look at light's behavior through a particle-driven lens and providing a fresh perspective on the universe.

In the following sections, we will continue to explore how the Vetter Theory provides explanations for other cosmic phenomena, from the fusion processes in stars to the behavior of cosmic background radiation. Together, these insights reinforce the proton-centered, particle-based view of the universe, showing how it can address a wide range of observed effects.

3.2 The Formation and Life Cycle of Stars

Proton Density and the Birth of Stars
In the Vetter Theory, the formation of stars begins with the accumulation of hydrogen—specifically, proton-rich hydrogen atoms—in dense molecular clouds. As these clouds grow denser, proton density reaches a critical level, where the strong force extends outward, producing the gravitational effect necessary for the matter to coalesce. This gravity, driven by the combined effect of proton clustering, acts as a catalyst for stellar formation, creating the conditions required for fusion.

- **Hydrogen Accumulation and Proton Clustering**: In interstellar space, regions with abundant hydrogen begin to pull together as proton density increases. The strong force, which drives gravitational attraction through proton density, causes hydrogen atoms to cluster, forming a gravitational well.
- **Gravitational Compression and Heating**: As proton-dense clouds pull matter inward, the increasing gravitational pressure raises temperatures within the core of the forming star. This compression generates the heat and pressure needed to ignite fusion, which sustains the star over its life cycle.

- **Critical Density for Fusion**: When proton density reaches a specific threshold, hydrogen atoms are compressed closely enough that their electrons and protons can interact at high energies, initiating the fusion process. This fusion process is not merely a nuclear reaction but involves the transformation of hydrogen atoms into new forms, driving the star's energy output.

This model positions proton density as the central factor in star formation, establishing gravity and fusion as emergent effects of hydrogen accumulation.

Fusion as a Proton-Electron Transformation Process

In traditional physics, fusion is explained as the process where atomic nuclei combine under extreme pressure and temperature, releasing energy and forming heavier elements. The Vetter Theory offers a distinct interpretation: fusion involves the interaction between a hydrogen atom's proton and electron, which combine to create a neutron, thus facilitating the formation of new elements.

- **Proton and Electron Fusion**: Within the intense environment of a star's core, electrons are pushed into close proximity with protons. Under these conditions, the proton and electron fuse to form a neutron, a process that underpins the formation of new, heavier elements.
- **Neutron Formation as the Key to Element Creation**: The fusion of protons and electrons into neutrons allows for the creation of heavier atomic nuclei. This process enables the formation of elements beyond hydrogen, with neutrons acting as spacers that prevent protons from repelling each other while maintaining the proton-driven strong force as the binding agent.
- **The Role of Neutrons in Heavier Elements**: As fusion progresses, the new neutrons allow protons to be spaced apart, creating stable atomic structures. In this way, the transformation of protons and electrons into neutrons facilitates the growth of more complex atomic structures within stars, enabling the synthesis of elements like helium, carbon, and beyond.

This fusion process highlights the unique transformation mechanism within the Vetter Theory, where proton-electron interactions facilitate neutron creation and elemental formation without requiring additional proton fusion.

Stellar Stability Through Proton Interactions

Once a star forms, it sustains itself by balancing the inward gravitational force with the outward pressure created by fusion. In the Vetter Theory, this balance is achieved through proton interactions, where the strong force maintains stability by regulating the density and behavior of particles within the star's core.

- **Pressure from Proton-Based Fusion**: The fusion process in the star's core generates immense energy, producing outward pressure that counteracts gravitational collapse. This balance between gravity and fusion keeps the star stable over billions of years.

- **Continuous Proton Interaction and Energy Release**: As protons interact with electrons and fuse into neutrons, they release energy in the form of light and radiation. This energy output is not only the source of the star's brightness but also a regulating mechanism that maintains the star's size and structure.
- **Hydrogen Depletion and Stellar Evolution**: Over time, hydrogen in the star's core is gradually converted into heavier elements. As hydrogen is depleted, the star undergoes changes, with fusion moving outward to new hydrogen-rich layers or, in larger stars, progressing to fuse heavier elements.

Stellar stability is thus an outcome of proton-based forces, where fusion serves as a dynamic counterbalance to gravitational pressure.

The Formation of Heavier Elements

In the later stages of a star's life, the core begins fusing elements beyond hydrogen. In the Vetter Theory, this process involves the continued transformation of hydrogen atoms into neutrons and their subsequent arrangement with protons, facilitated by the strong force.

- **Elemental Synthesis Beyond Helium**: As temperatures and pressures increase in the star's core, fusion extends to elements like carbon, oxygen, and iron. In this process, existing protons and electrons are converted to neutrons, enabling the assembly of nuclei with greater atomic mass.
- **Proton-Neutron Balance in Heavier Elements**: The presence of neutrons spaces protons apart, reducing electrostatic repulsion and creating stable structures. This spacing allows the strong force to bind larger numbers of protons in a nucleus without destabilizing it.
- **End-of-Life Fusion and Supernova Element Formation**: In massive stars, the fusion process ultimately produces elements up to iron. Once fusion can no longer sustain the star's gravitational pressure, the star collapses, often leading to a supernova. This explosive event releases neutrons and other particles, facilitating the formation of even heavier elements in the aftermath.

This neutron-based approach to element formation presents a unique view of stellar nucleosynthesis, where heavier elements are built through the fundamental transformation of protons and electrons.

A Proton-Centered View of Stellar Evolution and Death

As stars exhaust their fusion potential, they evolve into different end states based on their mass. In the Vetter Theory, the outcome of a star's life depends on proton density, quark pressure, and the presence of neutrons.

- **White Dwarfs and Proton Density Limits**: For stars with moderate mass, fusion ceases, and the star cools into a white dwarf. In these remnants, proton density reaches a stable threshold, where remaining protons and neutrons are tightly bound, preventing further collapse.

- **Neutron Stars as Neutron-Rich Remnants**: In more massive stars, the collapse leads to a neutron-dense structure where protons and electrons have fully combined into neutrons. This dense arrangement produces a highly stable configuration with intense gravitational pull, creating a neutron star.
- **Black Holes as Extreme Proton Densities**: The most massive stars collapse beyond the neutron star phase, reaching a point where proton density is so extreme that light cannot escape. In the Vetter Theory, black holes are not singularities but dense proton accumulations, creating intense gravitational fields due to the extreme proton concentration.

This view of stellar death and remnant formation underscores the role of proton and neutron density in determining a star's ultimate fate, aligning with the particle-based perspective of the Vetter Theory.

Proton Interactions as the Basis of Cosmic Element Creation

In the Vetter Theory, the formation and evolution of stars demonstrate how the proton serves as the master architect of cosmic structure. By driving fusion, gravitational formation, and the creation of heavier elements, proton interactions are shown to be fundamental to the cosmic life cycle.

- **A Continuous Cycle of Element Formation**: The process of fusion and neutron formation within stars allows for a continuous creation of elements, building complexity from the simplest hydrogen atoms to the heaviest naturally occurring elements.
- **Cosmic Recycling Through Stellar Evolution**: As stars evolve and eventually die, they release protons, neutrons, and other particles back into the universe, providing the raw materials for new stars and planetary systems. This recycling of matter supports the ongoing creation of complexity in the cosmos.
- **The Proton as the Foundation of Diversity in the Universe**: Through its ability to transform and interact within stars, the proton facilitates a universe rich in elemental diversity, laying the groundwork for planets, life, and complex structures.

This proton-centered view of stellar life and elemental creation positions the proton as the driving force behind both the physical and chemical diversity observed in the universe.

The Life Cycle of Stars: A Proton-Driven Narrative

In the Vetter Theory, stars are not simply nuclear reactors; they are proton-driven engines that power the universe's complexity. From their birth in hydrogen-dense clouds to their end as neutron stars or black holes, stars illustrate how the proton, through its interactions, drives cosmic evolution. This view redefines stellar processes and element formation, grounding them in the behavior of protons and the transformative fusion of protons and electrons into neutrons.

In the next section, we will explore how cosmic background radiation and quark pressure offer new perspectives on universal expansion, adding to the Vetter Theory's unified view of the cosmos.

3.3 Cosmic Background Radiation and Quark Pressure

A New Perspective on Cosmic Background Radiation
In conventional cosmology, cosmic background radiation (CBR) is interpreted as the residual heat from the Big Bang, a relic that provides insight into the universe's early state. The Vetter Theory, however, offers an alternative explanation: rather than being a one-time event's "afterglow," CBR is a result of ongoing quark interactions and pressure fluctuations in the quark matrix of space.

This interpretation presents CBR as a natural consequence of quark density adjustments in response to cycles of matter creation and annihilation. As matter is formed, destroyed, and cycled through the cosmos, quark pressure fluctuates, emitting low-energy radiation that permeates space. This radiation, unlike the Big Bang relic interpretation, is seen as a continuous process, shaped by quark dynamics rather than a single, ancient event.

Quark Density and the Emergence of Background Radiation
In the Vetter Theory, quarks are the essence of dark matter, filling the vacuum of space and forming a quark matrix that behaves much like a gas. When quark density fluctuates, as it does when matter cycles through creation and destruction, it produces low-energy radiation that we detect as CBR.

- **Quark Pressure as an Energy Source**: Quarks in the vacuum exert a form of pressure in response to their density. When quark density increases or decreases, quark pressure changes, releasing energy in the form of low-frequency electromagnetic radiation.
- **Uniform Radiation Across Space**: The quark matrix in space is vast and consistent, creating a uniform field of quark pressure that emits a nearly constant level of radiation. This consistency explains why CBR is observed uniformly in every direction, supporting a model of cosmic radiation that's naturally self-sustaining.
- **Fluctuations from Matter Cycling**: As black holes and other cosmic processes create and annihilate matter, the quark density in affected regions adjusts accordingly. These adjustments in quark pressure produce pulses of radiation, contributing to the overall background radiation detected across the universe.

This particle-based approach reframes CBR as a byproduct of quark dynamics, rooted in the constant, self-sustaining structure of the quark matrix rather than a singular event in cosmic history.

Quark Pressure and Cosmic Expansion
In addition to producing background radiation, quark pressure plays a fundamental role in driving cosmic expansion. In the Vetter Theory, this pressure acts as a force that naturally propels matter outward, creating the effect of dark energy. When quark density increases in response to matter creation or decreases due to annihilation, it produces an expansive force that contributes to the universe's growth.

- **Quark Pressure as Dark Energy**: As quark density in the vacuum increases, quark pressure generates an outward force that drives expansion, functioning as the source of dark energy. This expansive force doesn't require an unknown substance; it arises directly from the quark matrix's behavior.
- **Self-Regulating Cosmic Expansion**: Since quark density responds to cycles of matter creation and annihilation, the rate of cosmic expansion self-regulates based on particle interactions. Higher quark density produces greater quark pressure, causing a faster expansion rate, while lower density moderates expansion.
- **A Dynamic Balance Across the Universe**: Quark pressure provides a consistent, yet flexible, force that adjusts in response to matter fluctuations. This particle-based explanation offers a coherent view of dark energy, linking it directly to observable cosmic dynamics rather than hypothetical fields or energies.

Through the behavior of quark pressure, the Vetter Theory proposes a model where cosmic expansion is not a relic of an initial explosion but a continuous, self-balancing process driven by particle density.

Cosmic Background Radiation as a Constant Process

In the Vetter Theory, CBR is not the leftover heat from a singular creation event. Instead, it's a byproduct of an ongoing and balanced cosmic process, where quark density fluctuations emit radiation consistently over time. This approach aligns CBR with a universe in constant motion and self-renewal, shaped by particle interactions that maintain a uniform radiation field.

- **Background Radiation as a Steady Output**: Unlike a dissipating relic, CBR is generated continuously, as quark density fluctuates and emits low-energy radiation. This steady output means CBR is not tied to a specific point in time but is a fundamental, ongoing aspect of the universe's structure.
- **No Need for an Initial "Bang"**: By attributing CBR to quark dynamics, the Vetter Theory removes the need for a Big Bang or similar primordial event. Instead, it suggests that the universe's observable radiation is a result of natural processes that have always occurred and continue to shape the cosmos.
- **Testable Differences from Big Bang CBR**: If CBR originates from quark pressure rather than an initial explosion, subtle differences in temperature distribution and radiation patterns may exist. These could provide observable evidence for a particle-driven model, distinguishable from Big Bang-based predictions.

This view presents CBR as a stable, predictable process, grounded in quark interactions and unaffected by an ancient, singular event.

Observable Predictions and Potential Evidence

The Vetter Theory's view of CBR as quark-pressure radiation provides several unique observational predictions, distinct from Big Bang cosmology. These predictions offer avenues for testing the theory against current and future astronomical data.

- **Uniformity with Localized Variations**: Since CBR is generated continuously, there may be minor, localized variations in radiation density near active regions of matter cycling, such as around black holes or dense star-forming regions. This could result in slightly higher or lower radiation measurements in areas with high quark activity.
- **Radiation Stability Over Time**: The Vetter Theory predicts that CBR will remain consistent over cosmic time, as it's not the residue of an ancient event but an ongoing process. Unlike a gradually cooling relic, CBR in this model should exhibit stability and uniformity on a vast timescale.
- **Dark Energy Correlation with Quark Density**: If quark pressure drives cosmic expansion, then we should observe a correlation between quark density (and thus dark matter density) and expansion rates in different regions of the universe. Higher quark density should correspond to higher dark energy effects, creating a measurable relationship.

These predictions allow for direct comparisons between the Vetter Theory's approach to CBR and the Big Bang model, offering a way to validate or challenge the quark-based interpretation of background radiation.

Reinterpreting CBR and Cosmic Expansion as Quark-Driven Processes

In the Vetter Theory, CBR and cosmic expansion are interconnected phenomena, both driven by the behavior of quarks. This view creates a universe where radiation and expansion are natural consequences of particle interactions, rather than the aftermath of a distant explosion.

- **CBR as a Window into Particle Dynamics**: Rather than being a look back in time, CBR is a view of ongoing particle behavior in the present, revealing insights into how quarks and protons interact on a cosmic scale.
- **Dark Energy as a Function of Quark Pressure**: By linking quark pressure to dark energy, the Vetter Theory presents cosmic expansion as a stable, continuous process shaped by particle density. This provides a straightforward and testable explanation for the acceleration of the universe's expansion.
- **An Evolving, Self-Sustaining Cosmos**: This model of CBR and cosmic expansion supports the Vetter Theory's vision of a universe that sustains itself through natural particle interactions, offering a consistent framework for understanding the cosmos without needing external forces or events.

Cosmic Background Radiation and Quark Pressure: A New Model of the Universe's Structure

The Vetter Theory's interpretation of CBR and quark pressure redefines our understanding of cosmic background radiation and universal expansion. By grounding these phenomena in the behavior of quarks and their dynamic density, this approach provides a cohesive, particle-based explanation for two of cosmology's most enigmatic observations.

As we continue, we'll explore how this quark-based framework also sheds light on the nature of light and electromagnetic waves, further unifying the Vetter Theory's particle-centered view of the universe.

3.4 The Nature of Light and Electromagnetic Waves

Reinterpreting Light as a Product of Proton-Electron Interactions

In conventional physics, light and electromagnetic waves are explained as oscillations in the electromagnetic field, with photons as the force carriers. In the Vetter Theory, however, light and electromagnetic waves are direct outcomes of electron and proton interactions, driven by the strong force. This framework presents light not as a separate force or field but as an adaptable form of the strong force, shaped by the behavior of particles. In this view, light is generated by electron transitions and movements within the atom, particularly as electrons interact with protons. When electrons change positions or interact with surrounding protons, they produce oscillations that manifest as electromagnetic waves. This approach eliminates the need for a separate photon particle, positioning light as a natural outcome of atomic interactions.

Electron-Proton Dynamics as the Source of Light

The Vetter Theory proposes that the interaction between electrons and protons, modulated by the strong force, creates the conditions for light generation. As electrons respond to proton-based forces, they produce energy in the form of light and other electromagnetic waves.

- **Electron Transitions and Energy Emission**: When an electron moves between energy levels within an atom, it releases energy as light. This process is driven by the strong force's influence on electron positioning, creating a stable yet dynamic system for light production.
- **Electromagnetic Waves as Strong Force Variants**: Light is simply one expression of the strong force as it interacts with electron positioning. This view frames electromagnetic waves as adaptations of the strong force, responding to shifts in the electron's relationship with the proton.
- **Photon-Free Light**: In this model, light does not require a separate particle, like a photon, to carry it. Instead, light is an oscillatory effect of the proton-electron interaction, transmitted directly through the strong force's influence on surrounding particles.

By linking light to the strong force, the Vetter Theory unifies light production with the foundational particle interactions that define atomic structure.

Electromagnetic Waves Across the Spectrum

The electromagnetic spectrum, encompassing everything from radio waves to gamma rays, represents the diverse expressions of electron-proton interactions. In the Vetter Theory, each part of the spectrum corresponds to variations in electron behavior, producing different frequencies of electromagnetic waves without the need for an electromagnetic field.

- **Frequency as a Function of Electron Behavior**: The frequency of an electromagnetic wave is determined by the speed, intensity, and distance of the electron's movement in relation to the proton. Low-energy electron shifts produce radio waves, while high-energy transitions result in gamma rays.
- **Intensity and Wavelength Based on Strong Force Interactions**: The wavelength and intensity of light are direct outcomes of how the strong force influences electron-proton relationships. Longer wavelengths (like radio waves) are generated by minor shifts, while shorter, high-energy wavelengths (like X-rays) come from more significant transitions.
- **Electromagnetic Spectrum as a Continuum of Strong Force Manifestations**: Rather than viewing the electromagnetic spectrum as a separate field, the Vetter Theory presents it as a continuum of interactions within the strong force, modulated by electron behavior. This makes the entire spectrum a product of proton-based dynamics.

This approach frames the electromagnetic spectrum as a direct expression of particle interactions, simplifying the view of light and electromagnetic waves to a particle-centered model.

Observable Predictions for Light Behavior in High Proton-Density Regions

In regions of intense proton density, such as near black holes or neutron stars, the Vetter Theory predicts unique light behavior due to extreme interactions between protons and electrons. By understanding light as a strong force expression, we can make specific predictions about how it will behave in these environments.

- **Increased Intensity Near Proton-Dense Objects**: In high proton-density fields, such as near black holes, electron movements may be more energetic, producing higher-frequency electromagnetic waves. This could explain the intense X-rays and gamma rays observed near these objects without needing extreme gravitational fields.
- **Modified Light Path Due to Proton-Electron Interactions**: Light waves near massive objects should exhibit shifts due to the concentrated proton density, altering the path of light based on direct particle interactions. This can create lensing effects that align with observed gravitational lensing patterns.
- **Electromagnetic Spectrum Shifts in Extreme Fields**: Under high-proton density conditions, we would expect the electromagnetic spectrum to shift, with an abundance of high-energy waves like gamma rays and X-rays. This prediction aligns with observational data from active galactic nuclei and other high-energy cosmic phenomena.

These predictions provide observable evidence for the Vetter Theory's particle-driven approach to light, distinguishing it from traditional field-based models.

Electromagnetic Waves and Atomic Structure

In the Vetter Theory, the production and transmission of electromagnetic waves are closely tied to atomic structure. Each atom's configuration dictates how it generates and interacts with light, providing a direct link between atomic-scale dynamics and electromagnetic phenomena.

- **Atomic Energy Levels and Light Emission**: The arrangement of electrons around the proton in each atom determines how light is emitted. Electron transitions across energy levels are directly influenced by the proton-driven strong force, resulting in unique emission spectra for each element.
- **Chemical Bonds and Molecular Light Behavior**: In molecules, electrons interact across multiple atoms, producing complex electromagnetic behaviors. Covalent and ionic bonds modulate electron behavior, resulting in specific absorption and emission patterns that reveal the structure of the molecule.
- **Material-Specific Light Interactions**: Different materials exhibit unique responses to light based on their atomic structure. Metals, for example, conduct electromagnetic waves efficiently due to free-moving electrons, while insulators restrict electron movement, affecting their interaction with light.

This view connects the electromagnetic behavior of materials to the proton-electron dynamics within atoms, reinforcing a unified particle-based approach to understanding light.

Testable Predictions and Observable Implications

The Vetter Theory's interpretation of light and electromagnetic waves provides several testable predictions, which can be observed both in laboratory settings and in astrophysical observations.

- **Variability in Emission Spectrum in High-Proton Environments**: If light production is driven by electron-proton interactions, then variations in proton density should affect the electromagnetic spectrum. Near high-density objects like black holes, we would expect an enhanced emission of high-energy radiation (X-rays, gamma rays).
- **Absence of Photon Interaction in Atomic Transitions**: Traditional photon-based interactions predict specific quantum behavior in light emission and absorption. By contrast, the Vetter Theory's particle-based model suggests that light transitions should be more continuous and directly related to electron movement, potentially observable in controlled atomic experiments.
- **Predictable Spectrum Shifts Based on Proton Density**: By examining the spectrum emitted near various massive objects, researchers could correlate proton density with spectral intensity and energy. If these predictions hold, it would confirm the Vetter Theory's interpretation of light as a product of particle dynamics.

These predictions support the theory's view of light as an expression of atomic-scale particle behavior, offering avenues to test this particle-based approach against traditional field-based models.

Redefining Light as a Product of Particle Interactions

In the Vetter Theory, light is not an independent wave or field but a result of particle interactions, specifically the behavior of electrons in response to proton-driven forces. This approach presents light as a product of strong force dynamics, shaped by atomic structure and adaptable across a wide spectrum.

- **Light as an Observable Particle Interaction**: This view makes light a direct result of particle behavior, aligning it with the broader framework of proton-driven forces that govern atomic and cosmic scales.
- **A Consistent Model of Electromagnetic Behavior**: By eliminating the need for a separate photon particle, the Vetter Theory presents a streamlined view of light that ties it directly to atomic structure, allowing for a cohesive understanding of electromagnetic interactions.
- **Implications for Atomic and Astrophysical Research**: This particle-centered view of light could open up new research avenues in spectroscopy, particle physics, and astrophysics, offering a fresh perspective on the role of light in understanding atomic and cosmic structures.

The Nature of Light and Electromagnetic Waves: A Proton-Centered Approach

In the Vetter Theory, light is no longer a mysterious field but a direct result of particle dynamics. By understanding electromagnetic waves as manifestations of the strong force, this model unifies light with the foundational forces governing the universe, positioning it as an essential component of a proton-driven cosmos.

As we move forward, we'll see how this reinterpretation of light integrates with other cosmic phenomena, offering a cohesive view of the universe through a particle-based lens.

3.5 Galaxy Formation and Dark Matter

Proton Density and the Formation of Galactic Structures

In traditional astrophysics, galaxies are thought to form as gas and dark matter clump together, creating gravitational wells that attract stars and other matter. The Vetter Theory, however, redefines galaxy formation as a process driven solely by proton density. As hydrogen clouds accumulate and proton density increases, the strong force extends outward as gravity, creating a gravitational pull that draws matter inward, forming galaxies and binding them together without needing dark matter as an unseen mass.

- **Hydrogen Clouds as Gravitational Wells**: Large clouds of hydrogen, rich in protons, provide the foundation for galaxy formation. As proton density increases, gravitational effects emerge from the strong force, creating the gravitational fields necessary for pulling matter into cohesive structures.
- **Proton Clustering and Star Formation**: Within these hydrogen-dense clouds, regions of intense proton clustering lead to star formation. As hydrogen atoms coalesce under proton-driven gravity, stars ignite and begin to form the structure of a galaxy.

- **Gravitational Cohesion Across the Galaxy**: Proton density within galaxies generates gravity fields that prevent stars from drifting apart, establishing a gravitational well that defines the galaxy's shape and structure. This gravitational field, driven by proton interactions, allows galaxies to remain stable without invoking dark matter as an external force.

This model positions proton-based gravitational wells as the foundation of galaxy formation, with each galaxy's structure arising naturally from hydrogen accumulation and proton clustering.

Dark Matter as Free Quarks in Intergalactic Space

The Vetter Theory reinterprets dark matter as free quarks dispersed throughout the vacuum of space, forming a "quark matrix" that provides additional gravitational influence between galaxies. In this framework, quarks fill the intergalactic medium, creating a gravitational field that complements the gravitational wells formed by proton-dense galaxies.

- **Quark Matrix as a Gravitational Framework**: Free quarks act like a dispersed "gas" in the universe, filling the voids between galaxies. This quark matrix exerts a gravitational pull, influencing the motion and structure of galaxies without needing to be directly visible.
- **Additional Mass Effect from Quark Density**: The gravitational influence of the quark matrix provides the "extra mass" attributed to dark matter. As quark density increases in certain regions, it enhances gravitational attraction, creating observable effects on galaxy rotation and clustering.
- **A Natural Source of Gravitational Binding for Galactic Clusters**: By acting as a gravitational framework, the quark matrix allows galaxies to cluster together in groups and superclusters, creating structures that are gravitationally bound on a cosmic scale. This gravitational effect explains the coherence of galaxy clusters without needing dark matter as an exotic substance.

This interpretation of dark matter as free quarks positions the quark matrix as a particle-based explanation for the additional gravitational effects observed between galaxies.

Observable Effects on Galaxy Rotation and Structure

One of the main challenges in astrophysics is explaining why galaxies rotate as if they contain more mass than is visible. The Vetter Theory's view of dark matter as a quark matrix offers a particle-based explanation for these rotation patterns, suggesting that quarks in intergalactic space provide the additional gravitational pull required for galaxies to maintain their rotation speeds.

- **Rotational Curves and Quark Influence**: The gravitational pull from the quark matrix affects how galaxies rotate. By adding mass in the form of quarks distributed throughout space, the gravitational force experienced by outer stars aligns with the observed rotation curves, eliminating the need for dark matter halos.

- **Stabilizing Galaxy Structures**: The quark matrix creates a gravitational "buffer" around galaxies, which stabilizes their structures and prevents stars from dispersing. This quark-based buffer allows galaxies to maintain structural integrity even in regions far from the galactic core.
- **Cluster Behavior and Dark Matter Distribution**: In galaxy clusters, the quark matrix produces a collective gravitational effect, enhancing the cohesion of galaxies within clusters. Observational data on galaxy cluster dynamics can thus be explained through quark density rather than relying on hypothetical dark matter particles.

This interpretation offers a coherent explanation for galaxy rotation and stability, providing an observable framework based on quark density that could be tested against astronomical data.

Quark Density Fluctuations and Large-Scale Structure

In the Vetter Theory, variations in quark density in the quark matrix play a key role in shaping the universe's large-scale structure. As quark density fluctuates, it creates regions of stronger or weaker gravitational influence, leading to the formation of filaments, voids, and cosmic web patterns across the universe.

- **Filament and Void Formation Through Quark Density**: Dense regions of quarks attract galaxies, forming cosmic filaments, while lower-density quark regions produce voids. This distribution aligns with observed large-scale structures in the universe, showing that quark density shapes the cosmic web.
- **Cosmic Web as a Particle-Driven Structure**: The quark matrix provides a natural framework for the cosmic web, where quark density variations create the gravitational scaffolding for galaxies to align and cluster in filaments. This particle-based approach offers an intuitive explanation for the observed cosmic structure.
- **Dynamic Adjustment Through Matter Cycling**: As matter cycles through creation and annihilation in the universe, quark density adjusts accordingly, creating a dynamic balance in the cosmic web. These density adjustments contribute to the self-regulating balance of the universe's large-scale structure.

By positioning quarks as the gravitational scaffolding of the cosmos, the Vetter Theory provides a particle-driven explanation for the cosmic web and large-scale structure formation.

Observable Predictions for Dark Matter and Galactic Structure

The Vetter Theory's view of dark matter as a quark matrix provides several testable predictions regarding galaxy behavior, structure, and clustering. These predictions offer avenues to compare the quark-based model against traditional dark matter theories.

- **Rotational Symmetry Correlations with Quark Density**: If quark density drives galaxy rotation, then rotational symmetry should correlate with quark density patterns across different galaxies. Observing these correlations would provide evidence supporting a quark matrix rather than dark matter halos.

- **Gravitational Lensing in Intergalactic Quark Fields**: The quark matrix should influence gravitational lensing between galaxies, producing lensing effects consistent with a distributed particle field rather than a concentrated dark matter halo. Observing these lensing patterns could differentiate the Vetter Theory's model from conventional dark matter.
- **Large-Scale Structure Consistency**: The cosmic web's distribution and alignment should directly correlate with quark density, producing predictable patterns in galaxy clustering. By examining large-scale cosmic structures, researchers could validate the quark-driven interpretation of dark matter's gravitational effects.

These predictions provide a means of testing the Vetter Theory's view of dark matter as a quark matrix, offering a particle-based alternative to traditional dark matter theories.

A Particle-Based Model of Dark Matter and Galactic Formation

The Vetter Theory redefines galaxy formation and dark matter through proton density and quark interactions. By positioning free quarks as the source of additional gravitational effects and viewing galaxies as proton-driven structures, this approach offers a particle-centered framework that aligns with observed cosmic phenomena.

- **Dark Matter as a Quark Matrix**: This interpretation eliminates the need for exotic particles, grounding dark matter's gravitational effects in known particle interactions. Quarks in intergalactic space provide the "extra mass" effect without invoking unseen matter.
- **Galactic Formation as a Proton-Driven Process**: Galaxy formation arises naturally from proton density, with hydrogen clouds generating gravitational wells that draw matter into stable structures. This model provides a cohesive view of galaxy dynamics rooted in proton-based forces.
- **Unified View of Cosmic Structure**: By linking quark density to dark matter and galaxy formation to proton density, the Vetter Theory creates a unified model of the universe's structure, offering a cohesive, particle-driven explanation for large-scale cosmic phenomena.

Galaxy Formation and Dark Matter: Rethinking Cosmic Structure

In the Vetter Theory, galaxy formation and dark matter are particle-based phenomena, emerging from proton and quark interactions. This view provides a straightforward, testable model for understanding the universe's structure, offering an alternative to traditional dark matter interpretations and opening the door for new observational opportunities. In the next section, we'll delve into black holes and their role as dense proton accumulations, further exploring how proton density influences cosmic phenomena within the Vetter Theory.

3.6 Black Holes and Proton Density Thresholds

Redefining Black Holes as Dense Proton Accumulations

In traditional physics, black holes are described as singularities—points of infinite density that warp spacetime so intensely that nothing, not even light, can escape. The Vetter Theory offers a different perspective: black holes are not singularities but instead are regions of extreme proton density. In this view, black holes represent proton accumulations that reach a critical density, creating intense gravitational fields due to the strong force and eliminating the need for spacetime curvature.

In the Vetter Theory, black holes are simply dense configurations of protons clustered tightly together. As proton density increases beyond a certain threshold, the gravitational pull intensifies to the point where it overwhelms nearby particles, including photons. This proton-driven approach to black holes simplifies our understanding of these cosmic objects, focusing on particle density rather than abstract singularities.

Proton Density Thresholds and the Formation of Black Holes
The formation of a black hole, according to the Vetter Theory, occurs when a massive star collapses, compressing its protons to the point where their density exceeds a critical threshold. At this threshold, the strong force extends outward, creating a gravitational effect so powerful that it prevents any matter or light from escaping.

- **Collapse of Massive Stars**: When a star exhausts its fusion fuel, it collapses under its own gravitational pull. In high-mass stars, this collapse results in a concentration of protons, leading to an exceptionally dense core.
- **Reaching Critical Proton Density**: As protons continue to compress, their density increases to the point where the strong force produces extreme gravitational effects, pulling in all surrounding matter. At this threshold, the proton density becomes so high that light and matter are unable to escape.
- **No Need for a Singularity**: Unlike the traditional singularity model, this proton-density approach does not require infinite density. Instead, black holes are stable, finite clusters of protons with a density that produces a gravitational field strong enough to prevent escape.

This model redefines black holes as physical, dense objects rather than singularities, aligning with the Vetter Theory's particle-centered framework.

Electron Ejection and Light Interaction Near Black Holes
One unique aspect of the Vetter Theory's view of black holes is the ejection of electrons near regions of extreme proton density. As matter approaches the black hole, the immense proton density forces electrons to separate from protons, ejecting them outward at high speeds. This electron ejection has observable effects on light and surrounding space, contributing to the lensing and radiation patterns commonly associated with black holes.

- **Electron-Photon Interactions**: The high-speed ejected electrons near black holes interact with photons, bending their paths and creating the lensing effect often attributed to spacetime curvature. In this model, gravitational lensing is a particle interaction effect rather than a consequence of curved space.

- **Radiation and High-Energy Emissions**: As electrons accelerate away from the black hole, they emit radiation, particularly in the X-ray and gamma-ray spectra. This radiation is a direct result of particle interactions rather than gravitational waves, providing a straightforward explanation for high-energy emissions near black holes.
- **Observable Light Bending and Ejection Patterns**: The interaction between high-speed electrons and light creates bending patterns that can be observed with telescopes. These patterns offer a particle-based explanation for gravitational lensing, distinct from curvature-based interpretations.

By focusing on electron-proton dynamics, this view provides a tangible model for black hole behavior and offers testable predictions for light interactions near these objects.

Matter Cycling and Quark Pressure Adjustments Around Black Holes

In the Vetter Theory, black holes play an essential role in the universe's matter cycle. As they collect protons, black holes simultaneously influence quark density in the surrounding space. This matter cycling and quark pressure adjustment contribute to cosmic balance, ensuring that the universe maintains stability as matter is created and destroyed.

- **Proton Collection and Matter Recycling**: Black holes act as reservoirs, collecting protons from nearby matter and contributing to the ongoing cycle of matter in the universe. This accumulation process maintains a balanced matter distribution across space.
- **Quark Pressure as a Cosmic Balance Mechanism**: As black holes collect protons, the quark density in surrounding space adjusts to balance the gravitational field. This quark pressure provides a regulatory force that contributes to cosmic expansion and stability.
- **Matter Destruction and Reformation**: When protons reach a black hole, they do not disappear; instead, they cycle through stages of high-density quark adjustments, influencing cosmic structure. This constant cycling of matter supports the self-regulating framework of the universe as proposed in the Vetter Theory.

This cycling of matter through black holes presents a cohesive view of black holes as active contributors to the universe's dynamic balance, challenging the idea that they are simply regions of gravitational destruction.

Black Hole Behavior and Observable Predictions

The Vetter Theory's interpretation of black holes as proton-dense objects suggests several unique observational predictions, particularly regarding the behavior of light, matter, and radiation around these objects.

- **Lensing Patterns from Electron Interactions**: If lensing is caused by electron-photon interactions rather than spacetime curvature, the lensing patterns around black holes may exhibit asymmetries based on the electron ejection direction and speed, potentially observable with advanced telescopes.

- **Predictable High-Energy Radiation Near Proton-Dense Regions**: High-energy radiation, such as X-rays and gamma rays, should correlate with electron ejection patterns, providing a testable prediction for the Vetter Theory's model of black holes. Observing these patterns could support or challenge this particle-based interpretation.
- **Matter Accretion and Electron Behavior**: The Vetter Theory predicts that matter accreting around black holes will show distinctive electron-proton separation, leading to specific radiation and ejection patterns. This behavior could offer a new way to understand accretion disk properties and emissions.

These predictions allow for the empirical testing of the Vetter Theory's particle-based black hole model, offering a way to compare its predictions against conventional models.

Redefining Black Holes as Dynamic Proton-Dense Regions

In the Vetter Theory, black holes are not static regions of infinite density but dynamic proton-dense regions that influence their surroundings. By focusing on proton and electron interactions, this view creates a model of black holes as active participants in cosmic structure and balance.

- **A Finite-Density Model of Black Holes**: The Vetter Theory's approach removes the need for singularities, presenting black holes as dense yet finite regions, stabilized by proton clustering rather than abstract curvature.
- **Gravitational Effects as Particle-Driven Forces**: Gravity near black holes is explained through proton density, making it a particle-based effect that does not require spacetime distortion. This offers a practical, observable view of gravity around massive objects.
- **A Role in Cosmic Cycling and Stability**: By collecting protons and adjusting quark density, black holes contribute to the universe's self-regulating structure. This role aligns with the Vetter Theory's framework of a balanced, particle-driven cosmos.

This redefinition of black holes provides a consistent, particle-based explanation for their properties, positioning them as essential components of cosmic structure.

Black Holes and Proton Density Thresholds: A New Perspective

In the Vetter Theory, black holes are not mysterious points of infinite density but are natural outcomes of proton density thresholds. This particle-centered approach offers a fresh view of black holes as stable, active regions that shape cosmic dynamics, providing both gravitational balance and a continuous cycle of matter.

In the next section, we'll examine how these principles lead to testable predictions and observable implications across the cosmos, consolidating the Vetter Theory's particle-based approach to universal phenomena.

3.7 Testable Predictions and Observational Implications

A New Lens on Observable Phenomena

The Vetter Theory presents a particle-driven, proton-centered model of the universe, fundamentally reinterpreting how forces and structures arise on both atomic and cosmic scales. This section consolidates the theory's predictions, providing observational implications that could be tested with current or future technology. By exploring observable behaviors of light, gravity, cosmic radiation, and particle interactions, the Vetter Theory offers concrete avenues for scientific investigation.

These predictions aim to provide a testable framework for the theory, distinguishing its outcomes from conventional models and challenging existing assumptions in cosmology and particle physics.

Gravitational Lensing and Light Bending Predictions

In the Vetter Theory, gravitational lensing is a result of electron-photon interactions rather than spacetime curvature. This view produces specific predictions about how light will behave near massive objects like black holes and dense star clusters.

- **Asymmetrical Lensing Patterns**: If lensing results from electron-photon interactions, then the bending of light should exhibit asymmetries based on the direction and speed of electron ejections. Observations of gravitational lensing around black holes and dense galactic cores could reveal subtle variations not predicted by curvature-based models.
- **Intensity-Dependent Light Bending**: The Vetter Theory predicts that lensing strength may vary with proton and electron density, suggesting that regions with higher electron ejection rates, such as active galactic nuclei, will show stronger or more complex lensing effects.
- **Photon Redirection Differences**: Unlike gravitational lensing from spacetime curvature, this particle-based lensing would cause slight differences in the paths and arrival times of photons passing near dense objects, potentially observable with high-precision measurements.

These predictions could be tested by analyzing light behavior around known high-density areas, offering a way to evaluate the Vetter Theory's approach to light and gravity.

Galaxy Rotation and Quark Matrix Effects

The Vetter Theory reinterprets dark matter as a quark matrix, suggesting that quarks between galaxies create the gravitational effects necessary to explain galaxy rotation and clustering. This leads to specific predictions about galactic dynamics and dark matter distribution.

- **Rotation Curves Without Dark Matter Halos**: If galaxy rotation is driven by a quark matrix, then galaxies should display predictable rotation curves that align with quark density rather than requiring dark matter halos. Observations of galaxy rotation curves could reveal whether they correspond to predicted quark distributions.

- **Dark Matter Lensing Patterns Based on Quark Density**: The quark matrix should influence gravitational lensing patterns around and between galaxies. Observing these patterns could help distinguish between a quark-based dark matter model and a traditional dark matter particle model.
- **Large-Scale Structure Correlation with Quark Density**: The cosmic web, including galaxy filaments and voids, should correlate directly with quark density variations. If quark density affects large-scale cosmic structure, then clustering patterns and galaxy distribution could offer evidence supporting the Vetter Theory's interpretation.

These predictions provide observational criteria for testing whether quarks in space function as a gravitational matrix, potentially offering a new perspective on dark matter.

Cosmic Background Radiation and Quark-Driven Expansion

In the Vetter Theory, cosmic background radiation (CBR) is a result of ongoing quark interactions rather than a relic of the Big Bang. This theory suggests that CBR is emitted continuously through quark pressure adjustments, providing a new way to understand background radiation.

- **Uniformity with Localized Variations**: Since CBR arises from quark interactions, localized variations in quark density near high-density regions, such as around black holes, could lead to minor fluctuations in the CBR. Detecting these localized variations would support the theory's quark-driven model.
- **Stable CBR Intensity Over Time**: Unlike a Big Bang relic that would gradually dissipate, quark-generated CBR should exhibit stable intensity over cosmic time. Consistency in CBR measurements across vast time spans would align with the Vetter Theory's continuous radiation model.
- **Correlation with Cosmic Expansion and Dark Energy**: If quark pressure drives cosmic expansion, then regions with higher quark density should experience accelerated expansion. Observing a correlation between CBR patterns and expansion rates would provide evidence for the theory's quark-based dark energy model.

These predictions offer an alternative view of CBR, distinguishing it from relic-based interpretations and grounding it in continuous particle interactions.

Black Hole Behavior and Proton Density Thresholds

The Vetter Theory redefines black holes as dense proton accumulations rather than singularities, predicting distinct behaviors in the vicinity of these objects, particularly in terms of light, matter, and radiation interactions.

- **Unique Lensing Effects from Electron Ejections**: The theory predicts that electron-photon interactions will produce lensing patterns around black holes that vary based on electron ejection rates and directions. High-precision observations of lensing near black holes could reveal these asymmetrical patterns.

- **Predictable High-Energy Radiation from Electron Ejections**: As electrons are ejected from high-proton-density regions around black holes, they should produce X-rays and gamma rays. Observing these high-energy emissions, particularly their correlation with electron behavior, could validate the particle-driven view of black holes.
- **Matter Accretion and Electron-Proton Separation**: The Vetter Theory predicts that matter accreting around black holes will exhibit distinctive electron-proton separation, leading to observable emission patterns. This could provide a new way to understand accretion disk radiation, contrasting with traditional models.

These predictions offer a framework for testing the Vetter Theory's view of black holes as proton-dense regions, providing observable distinctions from singularity-based models.

Electromagnetic Wave Behavior in High-Density Regions

In the Vetter Theory, electromagnetic waves are a product of electron-proton interactions, governed by the strong force. This interpretation suggests testable predictions about light behavior in environments with intense proton density.

- **Spectrum Shifts in Proton-Dense Regions**: In areas with high proton density, the electromagnetic spectrum should shift, with increased emission of high-energy waves like gamma rays and X-rays. Observing these shifts near dense cosmic objects could provide evidence for the theory's particle-based light model.
- **Material-Specific Light Interaction Patterns**: Different materials should exhibit unique electromagnetic behaviors based on their atomic structure, as the strong force influences electron behavior in diverse ways. This could be tested through spectroscopy and light-matter interaction studies.
- **Absence of Photon Particles in Atomic Transitions**: By focusing on particle interactions, the Vetter Theory implies that light transitions should occur without separate photon particles, potentially observable in controlled atomic experiments. This absence could provide a distinguishing characteristic from photon-based interpretations.

These predictions create a new framework for exploring the nature of electromagnetic waves and their origins, positioning light as a product of particle dynamics rather than independent field oscillations.

Toward a New Framework for Testing Cosmic Phenomena

The Vetter Theory's testable predictions provide a cohesive, particle-based framework for exploring cosmic phenomena. By grounding forces and structures in proton and quark interactions, this model offers a straightforward approach to testing fundamental aspects of the universe, challenging traditional concepts and opening the door to new scientific insights.

- **Distinguishing Particle-Driven Effects from Field-Based Models**: The Vetter Theory's predictions create a basis for comparing particle-driven effects against field-based models, providing a way to evaluate their relative accuracy and consistency with observable data.
- **Expanding Astrophysical Research with Particle-Centered Observations**: From galaxy rotation and dark matter effects to cosmic background radiation and black hole dynamics, the theory provides new observational avenues, encouraging research that explores proton and quark interactions on a cosmic scale.
- **Reframing Universal Forces Through Observable Particle Interactions**: This framework unifies cosmic forces as manifestations of the strong force, linking atomic and cosmic scales in a single, testable model of the universe.

As we move forward, these predictions will guide experimental research and observational astronomy, offering a new perspective on the cosmos grounded in the behavior of protons, electrons, and quarks.

Chapter 4

4.1 Reframing Fundamental Forces: A Unified Particle-Driven Model

Unifying the Four Fundamental Forces

In traditional physics, the universe's fundamental forces—gravity, electromagnetism, the strong nuclear force, and the weak nuclear force—are treated as separate interactions, each with unique carriers and underlying mechanisms. The Vetter Theory, however, challenges this division, positing that all forces are ultimately variations of the strong force, generated through proton interactions. This unified particle-driven model not only simplifies our understanding of forces but also eliminates the need for distinct force carriers, positioning the proton as the source of all interactions.

By seeing gravity, electromagnetism, and the weak force as adaptive manifestations of the strong force, the Vetter Theory offers a cohesive framework that connects atomic and cosmic scales, potentially leading to a more streamlined view of the universe's mechanics.

Gravity as a Proton Density Effect

In the Vetter Theory, gravity is not an independent force but an emergent effect of proton density. When protons cluster densely, as in stars or black holes, the strong force extends outward as gravitational attraction, pulling other protons, atoms, and objects toward the dense cluster. This interpretation positions gravity as a byproduct of proton density rather than a fundamental, standalone force.

- **Gravity's Dependence on Proton Clustering**: As proton density increases in objects like stars and planets, gravity emerges naturally. This approach eliminates the need for gravitons or spacetime curvature, simplifying gravity to a particle-driven effect.

- **A Cohesive Gravity Model for Atomic and Cosmic Scales**: Unlike conventional theories, which struggle to reconcile gravity at subatomic and cosmic scales, the Vetter Theory's proton-centered gravity unites these scales under a single principle. This allows for a consistent interpretation of gravity across the universe.
- **Implications for Black Holes and Stellar Structures**: The theory suggests that gravity reaches its peak in black holes, where proton density reaches critical levels, creating intense gravitational pull. This proton-driven model removes the need for singularities, reframing black holes as dense, stable clusters.

This view of gravity as a density-driven extension of the strong force has profound implications for theoretical physics, particularly in the context of unified force models.

Electromagnetism as a Variant of the Strong Force

In conventional physics, electromagnetism is understood as the force governing charged particle interactions, mediated by photons. The Vetter Theory reinterprets electromagnetism as another expression of the strong force, shaped by the electron's position relative to the proton within the hydrogen atom. By viewing electromagnetism as an effect of proton-electron dynamics, the theory eliminates the need for separate photons, grounding electromagnetic interactions in the proton-driven strong force.

- **Electron Position and Field Creation**: In the Vetter Theory, electron positioning around the proton generates electromagnetic fields as a natural extension of the strong force. This interpretation simplifies light and electromagnetic wave generation, making them direct outcomes of particle interactions.
- **Magnetism as Coordinated Proton-Electron Movement**: Magnetism arises when electron movements align with proton density, creating a magnetic field that is a coordinated effect of proton-electron interaction. This model eliminates the need for an independent electromagnetic field, linking it instead to atomic behavior.
- **Eliminating the Photon as a Force Carrier**: By interpreting light and electromagnetic waves as extensions of the strong force, the Vetter Theory removes the need for photons, instead attributing light to direct particle interactions. This simplifies the Standard Model by removing the photon as an independent particle.

This interpretation aligns electromagnetism with gravity and the strong force, positioning it as another expression of proton-based interactions and contributing to a unified force framework.

The Weak Force as a Proton-Based Transformation Process

The weak nuclear force, traditionally understood as the interaction responsible for radioactive decay and mediated by W and Z bosons, is also reinterpreted within the Vetter Theory. Instead of requiring separate mediating particles, the weak force is seen as a transformation effect, driven by instability within specific proton-electron configurations. In this model, the weak force enables particle transformations through proton interactions, aligning it with the strong force.

- **Particle Decay as Proton-Electron Instability**: The weak force is seen as a natural process within unstable proton-electron configurations, where protons transform into neutrons or other particles. This transformation is driven by internal proton dynamics rather than by W and Z bosons.
- **A Simplified Model of Radioactivity**: By positioning the weak force as a variant of the strong force, the Vetter Theory presents radioactivity as a proton-driven process. This eliminates the need for bosons, attributing particle decay directly to electron-proton instability.
- **Consistency Across Atomic and Cosmic Scales**: The reinterpretation of the weak force aligns it with gravity and electromagnetism, allowing for a single framework to explain transformations across different scales.

This unified view of the weak force offers a new perspective on particle decay, linking it to the Vetter Theory's particle-driven model of universal forces.

Implications for Quantum Mechanics and Relativity

By reframing all forces as expressions of the strong force, the Vetter Theory provides a framework that challenges the need for quantum mechanics and general relativity as separate theories. This model's simplicity suggests that both quantum mechanics and relativity could be unified under a single set of principles based on particle interactions, particularly those involving protons.

- **Simplifying Quantum Mechanics with a Proton-Centered Model**: In quantum mechanics, the behavior of particles is often described through probabilistic wavefunctions and quantum fields. The Vetter Theory's particle-based model replaces these abstractions with direct proton and electron interactions, potentially providing a more deterministic view of particle behavior.
- **Eliminating the Need for Curved Spacetime in Relativity**: General relativity explains gravity through spacetime curvature, which requires a fundamentally different framework from quantum mechanics. By interpreting gravity as a proton density effect, the Vetter Theory removes the need for curved spacetime, potentially unifying both fields under a single particle-based model.
- **Unifying Atomic and Cosmological Scales**: A unified model based on proton interactions could bridge the gap between quantum mechanics and relativity, offering a single set of principles for understanding particle and cosmic phenomena.

This potential unification represents a significant paradigm shift, suggesting that the mysteries of quantum mechanics and relativity could be simplified through the lens of the Vetter Theory.

Impact on the Standard Model and Force Carrier Particles

In the Standard Model, force carrier particles—photons, W/Z bosons, gluons, and gravitons—mediate interactions between particles. The Vetter Theory proposes that these carriers are unnecessary, as all interactions emerge from proton-based forces. This streamlined approach has significant implications for the Standard Model, potentially eliminating the need for multiple particle classifications and redefining fundamental physics.

- **Removing Force Carriers from the Model**: By reinterpreting forces as proton-driven phenomena, the Vetter Theory discards the need for distinct force carriers. This could simplify particle classification, focusing on particles that directly contribute to atomic structure, like protons, neutrons, and electrons.
- **Redefining the Higgs Mechanism**: The Higgs boson, often described as the particle that gives mass, could be reinterpreted in a proton-centered framework. Instead of a separate Higgs mechanism, mass could be seen as a result of proton density and electron interactions, eliminating the need for the Higgs boson as an independent particle.
- **A Leaner, More Cohesive Standard Model**: By removing force carriers, the Standard Model would consist primarily of protons, neutrons, electrons, and quarks, simplifying the framework and aligning it with the Vetter Theory's unified force model.

This redefinition of the Standard Model presents a streamlined view of particle physics, focusing on proton-based interactions rather than independent force carriers.

Reframing Forces: Toward a Unified Model of Physics
The Vetter Theory's reinterpretation of forces as expressions of the strong force provides a simplified, cohesive view of the universe's interactions. By eliminating separate carriers and unifying all forces under a single principle, this model offers a new path forward in physics, potentially bridging the gaps between quantum mechanics, relativity, and the Standard Model.

- **A Single-Force Model for Cosmic and Atomic Interactions**: This unified view allows for consistent explanations across scales, from the behavior of particles within atoms to the gravitational dynamics of galaxies.
- **Implications for Future Research and Experimentation**: With fewer theoretical components, the Vetter Theory could open up new avenues in both theoretical and experimental physics, encouraging a re-evaluation of particle interactions and the fundamental nature of forces.
- **A Foundation for Simplified Physics Models**: By reducing the complexity of the Standard Model and theoretical frameworks, the Vetter Theory provides a streamlined approach to physics, focusing on particle dynamics and the adaptable nature of the strong force.

This unified, proton-centered model redefines the fundamental forces as variations of a single, adaptable force, providing a cohesive framework for understanding the universe through a particle-based lens.

4.2 Rethinking Dark Matter and Dark Energy in Cosmology

Quarks as the Foundation of Dark Matter

In conventional cosmology, dark matter is proposed as an unseen form of matter that interacts gravitationally but not electromagnetically, thus remaining invisible. The Vetter Theory presents an alternative explanation: dark matter is not an exotic particle but instead is composed of free quarks dispersed throughout intergalactic space. These quarks form a quark matrix that exerts gravitational influence across cosmic distances, binding galaxies, and stabilizing galactic clusters.

This quark matrix offers a straightforward, particle-based explanation for the gravitational effects attributed to dark matter, positioning quarks as the "invisible" mass responsible for observed galaxy rotation curves and the large-scale structure of the universe.

- **Quarks as Distributed Mass**: Free quarks fill the vacuum of space similarly to a gas, creating a gravitational "scaffold" that stabilizes galactic clusters and contributes to cosmic cohesion. This provides the additional gravitational pull observed in galaxies without the need for hypothetical dark matter particles.
- **Eliminating the Need for Exotic Particles**: By framing dark matter as quarks, the Vetter Theory simplifies cosmology, eliminating the need to hypothesize new, unobserved particles and instead relying on quark interactions to explain the "missing" mass in the universe.
- **A Predictable Gravitational Influence Across Cosmic Scales**: The quark matrix's density is responsible for the gravitational effects observed on galactic and intergalactic scales. This consistent gravitational framework offers a new path for predicting galaxy rotation curves and clustering patterns based on quark density.

This approach to dark matter simplifies our understanding of cosmic structures, unifying visible and "invisible" matter under a single particle-based model.

Quark Pressure as the Source of Dark Energy

In conventional cosmology, dark energy is a mysterious force driving the accelerated expansion of the universe. The Vetter Theory proposes a different perspective: dark energy is not a separate entity but rather an emergent effect of quark pressure in the intergalactic quark matrix. As quark density fluctuates in response to matter cycling through creation and annihilation, quark pressure increases, driving cosmic expansion.

By interpreting dark energy as quark pressure, the Vetter Theory offers a testable, particle-based model for cosmic expansion that eliminates the need for hypothetical fields or repulsive forces, grounding cosmic acceleration in observable particle interactions.

- **Quark Density and Expansive Force**: As quark density increases, quark pressure creates an outward force that contributes to the expansion of space. This pressure, driven by quark density fluctuations, acts as the source of cosmic acceleration.

- **Self-Regulating Cosmic Expansion**: The quark matrix adjusts in response to the cycling of matter, creating a self-regulating balance that maintains the universe's expansion at a stable rate. This view aligns cosmic acceleration with natural particle interactions rather than external forces.
- **Predictable Expansion Rate Based on Quark Density**: The rate of cosmic expansion can be directly correlated with quark density. As the quark matrix fluctuates, the pressure it exerts produces observable changes in the rate of expansion, offering a way to model dark energy through quark behavior.

This particle-driven approach to dark energy provides a cohesive explanation for cosmic acceleration, simplifying cosmology by attributing expansion to known particle interactions.

Impact on the Cosmic Timeline and Structure Formation

By redefining dark matter and dark energy as quark-based effects, the Vetter Theory introduces a new framework for understanding the universe's timeline, structure formation, and evolution. Viewing cosmic expansion as a result of quark pressure and galactic cohesion as a function of quark-based dark matter, this approach could reshape models of cosmic history and future.

- **Adjusting the Age and Evolution of the Universe**: If cosmic expansion is driven by quark density, the timeline for expansion could differ from traditional models based on dark energy. This would affect estimates of the universe's age, potentially offering new insights into its early formation stages.
- **Quark-Driven Structure Formation**: The Vetter Theory suggests that quark density influenced the formation of galaxies and cosmic filaments, potentially leading to a unique timeline for structure formation. This could provide a new perspective on galaxy and cluster evolution, aligning structure formation with quark behavior.
- **Stability and Coherence of Galactic Clusters**: By positioning quarks as the foundation of dark matter, this model offers a coherent explanation for the stability of galactic clusters, attributing their cohesion to the gravitational influence of quarks in intergalactic space. This unifying view enhances our understanding of cosmic structure across scales.

This quark-based approach provides a particle-centered view of the universe's structure and evolution, linking dark matter and dark energy to observable particle dynamics.

Observable Implications and Testable Predictions

The Vetter Theory's interpretation of dark matter as quarks and dark energy as quark pressure offers several testable predictions, many of which could be observed through cosmic surveys and galaxy rotation studies.

- **Rotation Curves Consistent with Quark Density**: The gravitational influence of quarks in the quark matrix should produce specific galaxy rotation curves. These curves would correlate with quark density distributions rather than with dark matter halos, providing a clear criterion for testing the theory.
- **Localized Variations in Cosmic Expansion**: If dark energy arises from quark pressure, then variations in quark density should cause slight differences in the rate of expansion across different regions. Observing these localized expansion variations would provide evidence for the theory's quark-based dark energy model.
- **Gravitational Lensing Based on Quark Distribution**: The quark matrix's gravitational influence should produce predictable lensing patterns, particularly in intergalactic space. These lensing effects, distinct from traditional dark matter models, could reveal the quark-based structure of dark matter.

These predictions align the Vetter Theory's model with observable cosmic data, offering a framework for testing its validity against conventional interpretations of dark matter and dark energy.

Implications for the Cosmic Web and Large-Scale Structure

The Vetter Theory's quark-based dark matter model also has implications for the large-scale structure of the universe, including the cosmic web's filaments, voids, and clusters. By attributing large-scale structure to variations in quark density, this model provides a particle-driven explanation for the organization of galaxies across cosmic distances.

- **Formation of Cosmic Filaments and Voids**: The quark matrix's density variations create gravitational "scaffolding" that forms filaments and voids, providing a framework for the cosmic web. This structure is a natural outcome of quark density fluctuations rather than the clustering of dark matter particles.
- **A Quark-Driven Cosmic Web**: By interpreting the cosmic web as a result of quark distribution, the Vetter Theory offers a new perspective on the universe's large-scale structure. The distribution of galaxies along filaments could be directly correlated with quark density variations, providing observable evidence for the theory.
- **Dynamic Adjustment of Large-Scale Structure**: As matter cycles through creation and destruction, quark density adjusts, influencing the stability and evolution of the cosmic web. This dynamic, particle-based structure aligns with the Vetter Theory's view of a self-regulating universe.

This particle-centered view of the cosmic web enhances our understanding of cosmic structure, linking large-scale organization to quark interactions.

A New Framework for Dark Matter and Dark Energy

The Vetter Theory's quark-based interpretation of dark matter and dark energy provides a simpler, cohesive model for cosmology. By viewing quarks as the source of additional gravitational effects and cosmic expansion, this approach offers a particle-driven framework that eliminates the need for exotic matter or repulsive forces, grounding cosmic structure and expansion in observable particle behavior.

- **A Cohesive, Particle-Based Cosmology**: This quark-based model unifies dark matter and dark energy, providing a straightforward explanation for cosmic cohesion and expansion through quark density and pressure. This alignment simplifies our understanding of the universe's structure and behavior.
- **Implications for Future Research**: The Vetter Theory's quark-driven model of dark matter and dark energy encourages new observational methods, such as quark density mapping and localized expansion studies, potentially guiding future cosmic surveys and galaxy rotation studies.
- **A Step Toward Unifying Cosmology and Particle Physics**: By linking cosmic phenomena to quark behavior, the Vetter Theory bridges the gap between cosmology and particle physics, suggesting that understanding the universe's large-scale structure may lie in understanding its smallest particles.

This quark-based reinterpretation of dark matter and dark energy offers a fresh perspective on the universe, positioning particle interactions as the fundamental drivers of cosmic behavior and structure.

4.3 Advancing Astrophysics and Galactic Dynamics

Redefining Black Holes and Stellar Remnants
In the Vetter Theory, black holes are not singularities but proton-dense regions formed when a massive star's core collapses, concentrating protons to the point where intense gravitational forces arise. This proton-centered model redefines our understanding of black holes, suggesting that they are stable, finite-density objects with behaviors shaped by particle interactions rather than spacetime distortions.

- **Black Holes as Dense Proton Accumulations**: By seeing black holes as dense clusters of protons, this model eliminates the need for singularities or infinite density. Instead, black holes are stable, finite objects where proton density reaches a critical level, creating intense gravitational fields.
- **Impact on Black Hole Mergers and Gravitational Waves**: In this model, black hole mergers are interactions between dense proton clusters, potentially producing distinct gravitational wave patterns based on proton density rather than spacetime curvature. Observations of gravitational waves from mergers could reveal differences in the predicted waveforms.
- **Predicting High-Energy Radiation Patterns**: Since black holes eject electrons while retaining protons, this model predicts specific high-energy emissions (such as X-rays and gamma rays) as a result of electron ejections. Observing these radiation patterns near black holes could offer insights into their particle-driven behavior.

This particle-based view of black holes offers a more stable, finite interpretation of their structure, challenging traditional concepts of singularities and offering testable predictions for gravitational waves and high-energy emissions.

Quark-Based Galaxy Formation and Dark Matter Dynamics
The Vetter Theory's quark-based model of dark matter suggests that free quarks in intergalactic space create a gravitational "scaffold" that aids galaxy formation and stabilizes cosmic structure. This particle-driven framework reinterprets galaxy formation and rotation, attributing these phenomena to the gravitational influence of quarks rather than dark matter halos.

- **Galaxy Rotation Curves from Quark Influence**: By viewing quarks as the source of additional gravitational effects, this model predicts galaxy rotation curves without the need for dark matter halos. The quark matrix provides the necessary gravitational pull, explaining why galaxies rotate as if they contain more mass than is visible.
- **Stabilizing Galactic Clusters**: The quark matrix's gravitational influence acts as a binding agent for galactic clusters, stabilizing them and preventing dispersal. This view attributes cluster cohesion to particle density rather than hypothesized dark matter particles.
- **Galactic Filaments and Cosmic Web Structure**: Quark density variations create gravitational fields that attract galaxies into filament structures, forming the cosmic web. This particle-based model offers a cohesive explanation for the cosmic web, directly linking quark density to observed large-scale structure.

By positioning quarks as the driving force behind galactic structure, this model offers new insights into galaxy formation and clustering, potentially guiding future dark matter mapping and galaxy rotation studies.

High-Energy Phenomena and Cosmic Background Radiation
In the Vetter Theory, high-energy phenomena and cosmic background radiation (CBR) are viewed as outcomes of quark and proton interactions rather than relics of an ancient explosion. This framework suggests alternative explanations for gamma-ray bursts, quasars, and the origins of CBR.

- **Gamma-Ray Bursts as Quark Density Fluctuations**: The Vetter Theory interprets gamma-ray bursts as regions where quark density temporarily spikes, releasing intense radiation as quarks interact. This model offers an alternative to traditional explanations involving massive star collapses or neutron star mergers.
- **Quasar Emissions as Electron-Quark Interactions**: Quasars, known for their intense radiation, are seen as dense regions where electron ejections and quark interactions produce high-energy emissions. This model suggests that quasars are powered by intense particle interactions, challenging the need for supermassive black holes as their sole source of energy.
- **CBR as an Ongoing Quark-Based Radiation**: Rather than being a remnant from the Big Bang, CBR is interpreted as continuous low-energy radiation from quark density fluctuations. This radiation fills space uniformly, consistent with CBR observations, and provides an ongoing background signal from quark interactions.

This interpretation of high-energy phenomena and CBR challenges conventional models, offering testable predictions and alternative explanations for observed cosmic radiation.

Implications for Observing High-Energy Astrophysics

The Vetter Theory's particle-driven approach to cosmic phenomena has significant implications for high-energy astrophysics, providing a new framework for interpreting radiation, emissions, and cosmic events. By attributing these phenomena to proton and quark interactions, this model suggests specific observational criteria that could distinguish it from conventional explanations.

- **Unique Gravitational Waveforms from Black Hole Mergers**: Observing gravitational waves from black hole mergers could reveal waveforms specific to proton-dense objects rather than singularities, providing a means of testing this model against traditional black hole theories.
- **Predictable Radiation Patterns Near Black Holes and Quasars**: The model predicts specific patterns in X-ray and gamma-ray emissions due to electron ejection from black holes and electron-quark interactions near quasars. Observing these patterns could offer insight into particle behavior near dense objects.
- **CBR Variations from Local Quark Density Changes**: Since CBR is seen as an ongoing radiation from quark density, localized variations in quark density could create minor fluctuations in CBR intensity. Observing these fluctuations would provide evidence supporting the Vetter Theory's interpretation of cosmic background radiation.

These predictions provide a new framework for high-energy astrophysics, guiding observational efforts toward verifying the Vetter Theory's particle-based model of cosmic radiation and structure.

Expanding Cosmic Surveys and Dark Matter Mapping

The Vetter Theory's view of dark matter as a quark matrix opens new avenues for cosmic surveys and dark matter mapping. By focusing on quark density rather than hypothesized dark matter particles, this approach suggests specific methods for studying galactic structures and cosmic evolution.

- **Quark Density Mapping to Predict Galaxy Clusters**: By mapping quark density variations, researchers could predict where galaxies are likely to cluster and form filaments. This could guide dark matter mapping efforts, providing a quark-based framework for understanding galaxy distribution.
- **Localized Expansion Variations as Quark Density Correlates**: If quark density influences cosmic expansion, then regional quark density variations should correlate with localized expansion rates. Studying these correlations could offer evidence for the quark-based dark energy model.

- **Structure Formation as a Function of Quark Distribution**: Observing the cosmic web's formation and the distribution of galaxies along filaments could reveal patterns aligned with quark density fluctuations, supporting the theory's model of large-scale structure.

These observational strategies offer a new approach to cosmic surveys, linking galactic structures and cosmic expansion to quark interactions and densities.

A New Paradigm for Astrophysics and Galactic Dynamics

The Vetter Theory's particle-driven model provides a fresh approach to understanding cosmic structure, high-energy phenomena, and radiation. By grounding astrophysical phenomena in quark and proton interactions, this model challenges conventional explanations and encourages new observational approaches.

- **A Particle-Based Model for High-Energy Astronomy**: The Vetter Theory's particle-driven approach provides a consistent explanation for gamma-ray bursts, quasars, and CBR, simplifying astrophysics by attributing cosmic phenomena to observable particle interactions.
- **Implications for Dark Matter and Cosmic Structure Studies**: This quark-based model offers a new path for dark matter and galaxy surveys, linking cosmic structure to quark density and redefining the cosmic web as a product of particle distributions.
- **Guiding Future Observations and Experiments**: The Vetter Theory's predictions provide a framework for future astronomical observations, encouraging studies that explore quark density, electron ejection, and proton-driven radiation across the universe.

This reimagined view of astrophysics, high-energy phenomena, and cosmic structure aligns with the Vetter Theory's unified model, offering a consistent, particle-based explanation for the dynamics of the cosmos.

4.4 Impact on Particle Physics and the Standard Model

A Single-Force Framework in Particle Physics

In the Vetter Theory, all forces—gravity, electromagnetism, the weak force, and the strong force—are variations of a single proton-driven force. This approach eliminates the need for separate mediators or force carriers, proposing that all interactions stem from proton dynamics. This unification of forces could have profound implications for particle physics, particularly in simplifying the complexity of the Standard Model.

- **All Forces as Extensions of the Strong Force**: By framing gravity, electromagnetism, and the weak force as expressions of the strong force, this model suggests that proton interactions are the sole source of all particle interactions. This replaces the need for independent force frameworks with a single, unified principle.

- **Simplifying Particle Interaction Models**: The Vetter Theory's proton-centered model could simplify our understanding of particle interactions, reducing them to expressions of proton-electron dynamics. This approach could enable more predictable models, eliminating the need for complex quantum field theory explanations for each force.
- **Unified Behavior Across Scales**: Unlike traditional particle physics, which separates quantum and gravitational interactions, the Vetter Theory allows for a unified model that applies consistently across atomic and cosmic scales, offering a new perspective for both high-energy particle experiments and gravitational studies.

By unifying forces under proton interactions, the Vetter Theory offers a simplified model that could reshape how particle physics interprets fundamental forces.

Removing Force Carrier Particles from the Standard Model

In the Standard Model, force carrier particles (such as photons, W/Z bosons, and gluons) are necessary to mediate interactions between particles. The Vetter Theory, however, removes the need for these carriers, interpreting all particle interactions as direct outcomes of the strong force emanating from proton dynamics. This shift could streamline the Standard Model, focusing only on particles with distinct structural roles, like protons, neutrons, and electrons.

- **No Need for Photons in Electromagnetic Interactions**: In this model, photons are not required to mediate electromagnetic interactions; instead, light and electromagnetic waves are seen as effects of electron-proton dynamics. This would remove photons from the Standard Model, simplifying the model and eliminating the need for quantized light particles.
- **Redefining the Role of W and Z Bosons**: Since the weak force is interpreted as an instability within specific proton-electron configurations, W and Z bosons are no longer necessary for mediating weak interactions. This reinterpretation reduces the need for complex boson interactions, focusing instead on electron-proton behavior.
- **Gluons Replaced by Direct Quark Interactions**: Rather than relying on gluons to hold quarks together, the Vetter Theory suggests that quark interactions are stabilized through the strong force in a direct manner. This shift simplifies the structure of the atomic nucleus by removing the need for gluons as intermediary particles.

This elimination of force carrier particles could streamline the Standard Model, making it less complex and more focused on primary particles and their direct interactions.

Reinterpreting the Higgs Mechanism and Particle Mass

The Higgs boson, which in the Standard Model is believed to give particles mass, may be reinterpreted within the Vetter Theory's proton-centered framework. In this view, mass is a function of proton density and electron dynamics rather than the Higgs mechanism. This shift redefines the role of mass in particle physics, positioning it as an intrinsic property of particle interactions.

- **Mass as a Result of Proton Density**: In the Vetter Theory, the mass of particles is determined by proton density and quark distribution rather than by a separate field or particle. This approach aligns mass with proton interactions, eliminating the need for the Higgs field.
- **Simplifying Mass-Energy Relationships**: By linking mass directly to proton density, this model provides a more straightforward explanation for mass-energy equivalence, focusing on quark and proton dynamics. This perspective could simplify high-energy physics calculations and interpretations.
- **Eliminating the Higgs Boson from the Standard Model**: If mass is an intrinsic property of proton and quark interactions, then the Higgs boson may not be necessary. This redefinition removes the Higgs boson as a fundamental particle, reducing the number of elementary particles in the Standard Model.

This reinterpretation of mass challenges the current understanding of the Higgs mechanism, suggesting a simpler approach to particle mass that aligns with a proton-driven framework.

Redefining Particle Interactions and Decay Processes

In the Vetter Theory, particle decay and transformations are driven by proton-electron instability rather than mediated by separate bosons. This approach simplifies particle decay by attributing it to interactions within specific atomic configurations, providing a particle-centered view of transformation processes.

- **Instability-Driven Particle Decay**: Decay processes, such as beta decay, are seen as outcomes of instability within proton-electron configurations rather than the effect of W bosons. This simplifies decay explanations, focusing on particle dynamics rather than intermediaries.
- **A Deterministic Model for Particle Decay**: By attributing decay to specific configurations, the Vetter Theory offers a more deterministic view of particle decay, potentially allowing for predictions based on particle arrangement rather than quantum probabilities.
- **Neutron Formation Through Proton-Electron Fusion**: This model suggests that when protons and electrons combine, they form neutrons, facilitating the creation of heavier elements. This simplifies our understanding of neutron formation and enables a proton-centered explanation for atomic structure.

This approach to particle interactions and decay could lead to new experimental predictions and guide research in particle accelerators, potentially refining how decay processes are studied and understood.

Implications for Experimental Physics and Particle Accelerators

The Vetter Theory's redefinition of forces and particle interactions has implications for experimental physics, particularly in the context of particle accelerators. By eliminating force carriers and simplifying decay processes, this model offers new opportunities for testing particle behavior and mass generation without complex mediators.

- **New Experimental Approaches to Particle Decay**: Particle accelerators could test the theory's predictions regarding decay processes by observing configurations that lead to decay without boson mediation. This could refine the understanding of decay and transformation in subatomic particles.
- **High-Energy Tests of Proton-Centered Mass Generation**: By testing mass generation as a function of proton density, experiments could determine whether mass is indeed intrinsic to proton interactions rather than reliant on a Higgs field. High-energy experiments could provide insights into this mass origin theory.
- **Potential Absence of Force Carriers in Particle Interactions**: The Vetter Theory predicts that particles interact directly without mediating bosons. By testing this, accelerators could potentially observe interaction patterns consistent with a proton-driven model, helping to validate or challenge the need for boson carriers.

These experimental predictions provide a new avenue for particle physics research, encouraging studies that focus on direct particle interactions and mass as a proton-driven effect.

Toward a Leaner, Unified Standard Model

The Vetter Theory's particle-centered model proposes a simpler Standard Model, where all forces are extensions of the strong force, and interactions occur directly between particles. This model eliminates boson carriers and redefines mass, offering a streamlined framework that aligns with the theory's unified force model.

- **A Minimalist Standard Model**: By removing photons, W/Z bosons, gluons, and the Higgs boson, this model reduces the Standard Model's complexity. The remaining framework focuses on particles with primary structural roles, such as protons, neutrons, electrons, and quarks.
- **Direct Particle Interactions as the Foundation of Physics**: This proton-driven model positions direct interactions as the basis of all forces and transformations, simplifying particle classifications and interactions within a single framework.
- **Consistent Explanation Across Scales**: By unifying forces, the Vetter Theory's interpretation provides a consistent explanation that bridges subatomic and cosmic scales, aligning particle physics with astrophysics under a single theoretical model.

This streamlined Standard Model presents a new way of understanding fundamental physics, focusing on proton-based interactions to unify and simplify the universe's forces.

Reimagining Particle Physics: A Unified Proton-Driven Model

The Vetter Theory's impact on particle physics provides a fresh, simplified framework for interpreting fundamental interactions. By grounding all forces in the strong force and removing complex force carriers, this model offers a leaner, more cohesive Standard Model with wide-reaching implications.

- **A Simplified View of Particle Interactions**: By unifying forces and removing mediating bosons, the Vetter Theory offers a straightforward interpretation of particle interactions, challenging current models and encouraging experimental verification.
- **Implications for Theoretical and Experimental Research**: This model invites new research approaches in particle accelerators, dark matter studies, and gravitational wave research, guiding experiments that test proton-based mass and interaction theories.
- **A Proton-Centered Standard Model**: By focusing on direct proton interactions, the Vetter Theory presents a cohesive, particle-driven view of the universe, offering a new perspective on forces, mass, and particle dynamics that could reshape the future of physics.

This section concludes by positioning the Vetter Theory as a potential unifying model for particle physics, presenting testable predictions that challenge the foundations of the current Standard Model.

4.5 Applications in Technology and Engineering

Proton-Based Energy Generation

The Vetter Theory's interpretation of protons as the source of all forces introduces a new perspective on energy generation. By understanding energy as a direct outcome of proton-electron interactions, this model opens up potential applications in creating efficient, high-output energy sources. With the elimination of force carriers and a focus on particle dynamics, energy technologies could benefit from more streamlined and powerful mechanisms.

- **Fusion Reimagined as Proton-Electron Fusion**: Fusion energy technology, which traditionally focuses on combining atomic nuclei to release energy, could be reimagined based on the Vetter Theory. Fusion, in this model, involves the combination of protons and electrons to form neutrons, producing energy more directly and potentially in a controlled manner.
- **Energy from Controlled Matter Cycling**: By leveraging the constant creation and destruction of matter through proton and quark dynamics, new technologies could cycle particles in a way that continuously generates energy. This proton-centered approach might enable sustainable energy sources that rely on particle interactions rather than complex reactor designs.

- **Efficient Proton-Driven Energy Systems**: Proton interactions could lead to more efficient energy systems by reducing energy losses associated with force carriers. By focusing on direct particle interactions, energy technologies could become more compact and high-yield, particularly in applications where space and efficiency are critical, such as in spacecraft propulsion.

This redefined approach to energy generation has the potential to revolutionize the energy sector, making it possible to harness energy from direct particle interactions with fewer intermediate steps.

Advances in Materials Science

The Vetter Theory's focus on proton and quark interactions could lead to significant advances in materials science. By understanding materials at a proton-centered level, researchers could develop materials with unique electromagnetic, thermal, and structural properties, enhancing fields such as electronics, construction, and nanotechnology.

- **Electromagnetically Responsive Materials**: By controlling proton density and electron positioning within materials, it may be possible to create materials that respond dynamically to electromagnetic fields. These materials could change their properties on demand, potentially useful in applications like smart fabrics or adaptive building materials.
- **Superconductors with Enhanced Stability**: Proton-based materials could lead to the development of superconductors that operate at higher temperatures or under varying conditions, opening up applications in energy transmission, magnetic levitation, and medical imaging.
- **Radiation-Resistant and High-Durability Materials**: By manipulating proton and quark interactions, materials could be engineered to withstand extreme environments, such as high-radiation or high-heat settings. This would be especially useful in aerospace, nuclear reactors, and deep-sea exploration, where durability is essential.

This proton-centered approach to materials science has the potential to produce new classes of materials with enhanced capabilities, tailoring their properties for specific technological needs.

Technological Implications of Electromagnetic Wave Behavior

The Vetter Theory's interpretation of light and electromagnetic waves as manifestations of the strong force offers new possibilities for manipulating electromagnetic fields. By treating electromagnetic waves as extensions of proton-driven interactions, this model could lead to advances in telecommunications, imaging, and even computing.

- **Enhanced Telecommunications and Signal Clarity**: By understanding electromagnetic waves as proton-electron interactions, engineers could design systems that manipulate wave frequencies more precisely, potentially reducing signal loss and improving clarity in telecommunications. This could enhance both wireless and fiber-optic communications.

- **Advanced Imaging and Spectroscopy**: With the ability to control the behavior of electromagnetic waves, the Vetter Theory's framework could improve imaging and spectroscopy technologies, enabling high-resolution imaging in fields such as medicine, security, and remote sensing.
- **Quantum Computing and Data Processing**: Proton-based wave manipulation could lead to breakthroughs in quantum computing, where precise control over particle interactions is key. This approach could enable faster, more accurate processing, with applications ranging from cryptography to complex scientific modeling.

By focusing on the strong force in electromagnetic waves, the Vetter Theory provides a new way to enhance technologies reliant on electromagnetic interactions, broadening the scope for innovation in multiple fields.

Applications in Nanotechnology and Molecular Engineering

The Vetter Theory's proton-based approach to atomic and molecular interactions opens doors for advancements in nanotechnology and molecular engineering. By controlling proton and quark density at microscopic levels, scientists could develop molecular structures with highly specific properties, useful in medicine, environmental science, and computing.

- **Precision Molecular Assembly**: Understanding molecular bonds and interactions as proton-based could enable highly controlled molecular assembly, potentially useful for drug delivery systems that target specific cells in the body.
- **Self-Repairing and Self-Organizing Nanostructures**: By utilizing proton interactions, materials could be designed to self-repair or self-organize at the molecular level. This has applications in electronics, where circuits could self-heal, or in medical devices that adapt to changing conditions within the body.
- **High-Efficiency Molecular Sensors**: Proton-driven nanostructures could enhance the sensitivity of molecular sensors, allowing for the detection of trace elements or specific biomolecules. This could improve environmental monitoring, diagnostics, and biotechnological research.

These applications could push the boundaries of what is possible in nanotechnology, allowing scientists to leverage proton interactions for precise control at the molecular level.

Potential for New Propulsion Systems and Space Exploration Technologies

In the Vetter Theory, the proton-driven model of forces introduces new possibilities for propulsion and space exploration. By utilizing proton interactions as a stable energy source and focusing on high-efficiency, low-loss mechanisms, this model could pave the way for advanced propulsion systems and energy-efficient space technology.

- **Proton-Driven Ion Propulsion**: Ion propulsion, traditionally dependent on accelerating ions, could be reimagined using proton interactions to produce thrust. This approach might offer higher thrust-to-power ratios, enabling faster travel with lower fuel requirements.
- **Energy-Efficient Spacecraft Power**: With a proton-centered energy source, spacecraft could harness energy through direct particle interactions, reducing dependency on solar power or traditional fuel sources. This could enable long-duration missions with consistent energy output.
- **Materials for Extreme Space Environments**: Proton-based materials could be engineered for extreme conditions, such as radiation and temperature fluctuations, commonly encountered in space. These materials would enhance the durability of spacecraft, satellite, and planetary exploration equipment.

This approach to space technology introduces the potential for high-efficiency, durable systems, essential for advanced exploration and long-duration space missions.

Revolutionizing Computing and Data Processing

The Vetter Theory's simplified understanding of electromagnetic waves and particle interactions could have significant implications for computing. By eliminating the need for force carriers and focusing on proton-driven interactions, computing systems could be optimized for speed and efficiency, particularly in quantum computing and data processing.

- **Proton-Based Quantum Computing Models**: By interpreting quantum phenomena as proton-driven effects, quantum computing could advance with new models that focus on stable, direct interactions rather than entanglement. This could streamline quantum computing processes, making them more reliable and accessible.
- **High-Speed Data Transmission**: The Vetter Theory's electromagnetic wave interpretation could improve data transmission rates, particularly in fiber-optic and wireless networks, by reducing interference and signal degradation.
- **Energy-Efficient Processing Units**: By eliminating complex force interactions, processors could be designed to perform calculations with less energy loss, improving overall computational efficiency and reducing heat output in high-performance processors.

This particle-centered view of computing opens up possibilities for faster, more energy-efficient systems, with broad applications in scientific research, artificial intelligence, and large-scale data analysis.

A Technological Paradigm Shift: Innovations Rooted in Proton Dynamics

The Vetter Theory's proton-centered model suggests a new path for technological advancement, where innovations are guided by direct particle interactions rather than complex intermediary forces. By grounding technology in particle dynamics, this framework could lead to highly efficient, durable, and responsive systems across various fields.

- **Energy and Power Generation**: Proton-based energy generation systems could lead to sustainable power solutions that are both compact and efficient, benefiting everything from small electronics to large-scale power grids.
- **Enhanced Material Properties**: Materials engineered with a proton-centered understanding could become the foundation of next-generation electronics, construction, and medical devices.
- **Space Exploration and Propulsion**: By focusing on particle interactions, the Vetter Theory offers new propulsion and durability solutions, enabling advancements in space technology and exploration.

This section positions the Vetter Theory as a catalyst for technological innovation, offering practical applications across diverse fields, all grounded in a proton-centered understanding of forces and particle interactions.

4.6 Theoretical and Experimental Frontiers

Challenges in Testing the Vetter Theory
The Vetter Theory introduces a radical shift in how we interpret fundamental forces, cosmic phenomena, and particle interactions. As with any novel theory, validating these predictions poses unique challenges, particularly due to the need for high-precision data on particle interactions, proton density, quark behavior, and cosmic radiation. Addressing these challenges requires the development of new experimental methods and instruments capable of capturing these particle-centered phenomena.
- **High-Precision Particle Interaction Measurements**: Since the Vetter Theory relies on proton and quark dynamics, testing its predictions about particle decay, force interactions, and mass generation demands high-precision measurements of proton density and quark interactions. Current particle accelerators could be adapted to explore these dynamics, but additional refinements in detector sensitivity may be needed.
- **Localized Gravitational and Electromagnetic Observations**: Observing the Vetter Theory's predictions for gravitational lensing, dark matter distribution, and electromagnetic wave behavior requires highly sensitive telescopes and detectors. Instruments capable of capturing subtle variations in light and radiation patterns near massive objects could reveal new insights into electron-photon interactions and quark density.
- **Overcoming Inertial Assumptions of Traditional Physics**: The Vetter Theory challenges deeply rooted ideas like field theory and spacetime curvature, so overcoming the theoretical inertia of traditional models may also be a challenge. This theory demands a shift in perspective, encouraging physicists to focus on particle-driven explanations, particularly in fundamental forces.

These challenges highlight the need for both experimental innovation and theoretical flexibility, providing an impetus for developing novel techniques and refined instruments in astrophysics and particle physics.

Opportunities for Collaborative Research in Astrophysics and Particle Physics

Testing the Vetter Theory's predictions requires collaborative research efforts, particularly in areas where astrophysics and particle physics intersect. By working together, scientists across disciplines can design experiments and observations that target the theory's unique predictions, advancing a new era of interdisciplinary research.

- **Galactic Rotation and Dark Matter Studies**: Astrophysicists and particle physicists could collaborate to map quark density in galaxy clusters, testing the theory's prediction that quarks create the gravitational effect attributed to dark matter. This collaboration could involve both observational telescopes and dark matter simulation models, cross-referencing predicted quark densities with actual galaxy rotation curves.
- **Gravitational Waveform Analysis in High-Energy Environments**: Gravitational wave researchers and particle physicists could jointly analyze waveform data from black hole and neutron star mergers, looking for waveform signatures that align with proton-dense objects rather than singularities. This approach could test the theory's unique gravitational predictions against real-world observations.
- **Experimental Tests of Proton-Based Mass Generation**: Particle physicists could work with theorists to develop experiments that measure mass as a function of proton density rather than a Higgs field. High-energy experiments in particle accelerators could test whether mass originates from proton interactions alone, offering a direct test of the theory's stance on mass generation.

These collaborative efforts provide a pathway for validating the Vetter Theory's predictions, encouraging research that spans the full range of cosmic and atomic phenomena.

New Experimental Models and Technologies Based on the Vetter Theory

The Vetter Theory's principles offer new frameworks for designing experiments and technologies that explore fundamental forces, particle interactions, and cosmic expansion. These models could refine current techniques and open new frontiers in high-energy physics, gravitational studies, and cosmology.

- **Proton Density Measurement Techniques**: Since the Vetter Theory emphasizes proton density as a source of gravity, developing proton density measurement techniques could enable more accurate studies of gravity and mass. This could involve novel detectors or sensors capable of measuring variations in proton density in dense regions, such as stars and galaxies.
- **Quark Distribution Mapping**: Developing technologies to map quark distribution in space would allow researchers to visualize and quantify the quark matrix that serves as dark matter. Instruments designed to measure quark density could offer insights into galactic clustering and gravitational fields across the cosmic web.

- **Advanced Gravitational Wave Detectors for Non-Singularity Models**: Building gravitational wave detectors that analyze waveforms from the perspective of dense proton objects rather than singularities could reveal new patterns and provide empirical support for the Vetter Theory. Such detectors would be designed to capture nuances in gravitational waveforms that align with the theory's predictions.

These experimental models provide a foundation for validating the theory's unique claims, encouraging innovations that refine current measurement capabilities and support new areas of research.

Developing Theoretical Models Inspired by the Vetter Theory

In addition to experimental approaches, the Vetter Theory offers a foundation for developing new theoretical models that apply its principles to unexplained phenomena. By focusing on particle interactions and density-driven forces, theorists can explore novel explanations for phenomena that traditional models struggle to address.

- **Unified Field Theories Without Separate Force Carriers**: The Vetter Theory's single-force model provides a basis for developing unified field theories that do not rely on independent force carriers, such as photons or gluons. Theorists could explore mathematical frameworks that define all forces as proton-driven effects, offering a more cohesive and simplified view of particle physics.
- **Alternative Explanations for High-Energy Cosmic Events**: By reinterpreting phenomena like gamma-ray bursts and quasar emissions as outcomes of quark density fluctuations, theorists could create models that predict high-energy events based on proton and quark interactions alone. These models would provide an alternative to star collapse theories and dark matter particle annihilation.
- **Revisiting Quantum Mechanics Through Proton Interactions**: The Vetter Theory's deterministic view of particle interactions challenges quantum mechanics' probabilistic framework. By reinterpreting quantum phenomena as proton-driven effects, theorists could develop models that predict outcomes with more certainty, potentially redefining uncertainty principles and probabilistic interpretations.

These theoretical explorations could lead to a new generation of models in physics, guided by the Vetter Theory's principles and providing alternative explanations for both known and unknown phenomena.

Guiding Experimental and Observational Astronomy

The Vetter Theory's particle-centered approach also provides new criteria for observational astronomy, guiding researchers toward novel observational methods and tests that could validate or refute its predictions.

- **Cosmic Background Radiation (CBR) as Quark-Based Radiation**: Astronomers could observe CBR patterns to detect minor fluctuations in quark density across the universe. Localized variations in CBR intensity would support the theory's prediction that CBR is an ongoing radiation from quark interactions, offering a testable alternative to Big Bang relic interpretations.
- **Light and Gravitational Lensing Studies Near Massive Objects**: Observing light bending patterns near massive objects like black holes and neutron stars could reveal asymmetries in gravitational lensing caused by electron-photon interactions rather than spacetime curvature. These observations could validate the theory's particle-based lensing model.
- **Localized Expansion Variations and Dark Energy Tests**: If cosmic expansion is driven by quark pressure, then regions of varying quark density should exhibit slight differences in expansion rates. By measuring these localized expansion variations, astronomers could determine whether quark pressure contributes to dark energy effects, offering a direct test of the theory's view on cosmic expansion.

These observational opportunities guide astronomy toward specific tests for the Vetter Theory's predictions, encouraging research that aligns with the theory's particle-based cosmology.

A New Era for Physics and Cosmology: The Vetter Theory's Potential Legacy

The Vetter Theory's particle-driven model challenges established paradigms, introducing a unified force and quark-based explanations for dark matter, dark energy, and high-energy phenomena. By emphasizing proton and quark dynamics, this theory provides a foundation for new research avenues across physics and cosmology.

- **Innovating Instruments and Methods for Particle and Cosmic Studies**: The Vetter Theory encourages the development of instruments and observational methods focused on proton density, quark distribution, and particle interactions, guiding experimental physics and observational astronomy.
- **Bridging Gaps Between Physics Disciplines**: This particle-centered approach to universal forces offers a way to unify quantum mechanics, relativity, and the Standard Model, potentially bridging longstanding gaps between theoretical models in particle physics and astrophysics.
- **Laying the Groundwork for Future Paradigms**: By providing testable predictions and a simplified model of forces, the Vetter Theory invites physicists and astronomers to challenge existing concepts and explore a new paradigm, inspiring future generations to expand on these ideas.

This section concludes by positioning the Vetter Theory as a framework for pioneering research in both experimental and theoretical physics, encouraging a new approach to fundamental science grounded in particle interactions and simplified force dynamics.

Chapter 5

5.1 Reinterpreting Reality Through a Particle-Centered Lens

The Nature of Reality as Particle Interactions

In the Vetter Theory, reality is not an outcome of abstract fields or mysterious forces; rather, it is the result of tangible, consistent interactions among particles, specifically protons and quarks. This particle-centered view redefines reality as a cohesive and predictable system of forces that emerge directly from proton density and quark dynamics. By attributing all phenomena to these interactions, the theory proposes that the underlying mechanisms of existence are simple, stable, and observable.

- **Observable, Tangible Reality**: The Vetter Theory grounds reality in particles that can, in principle, be measured and observed, removing the need for elusive fields or unknown particles. In this view, reality is straightforward and accessible, encouraging a model that can be tested and refined based on direct observation and particle interactions.
- **Replacing Abstractions with Concrete Interactions**: Concepts like force carriers and field interactions are replaced with a single, unified force that manifests as particle interactions, making reality a sequence of tangible events rather than abstract, field-based phenomena.
- **A Reality Defined by Proton and Quark Behavior**: By positioning protons and quarks as the universe's building blocks, the Vetter Theory offers a cohesive, particle-driven model of existence. This approach suggests that understanding reality is a matter of comprehending particle behavior rather than navigating complex theoretical abstractions.

This particle-centered view presents reality as a structured, measurable phenomenon, encouraging a grounded approach to scientific inquiry and understanding.

A New Perspective on Time and Space

The Vetter Theory's dismissal of spacetime curvature and its focus on particle interactions introduces a unique perspective on time and space. Rather than viewing time as a fourth dimension bound to spatial dimensions, the theory presents time as a sequence of particle interactions, with space defined by proton and quark density. This shift challenges the traditional framework of spacetime, offering a simplified and unified interpretation.

- **Time as Sequential Particle Interactions**: In the Vetter Theory, time is not an independent dimension but a series of particle events. Each proton interaction represents a distinct moment, forming a continuous flow of events that constitutes time. This redefinition frames time as a tangible sequence rather than an abstract fourth dimension.

- **Space as Defined by Proton and Quark Density**: Space, similarly, is not an independent field but the distribution of protons and quarks. Dense regions form gravitational fields, while low-density areas create cosmic voids. This interpretation of space as a function of particle density simplifies cosmology, providing a particle-centered view of spatial structure.
- **Implications for Temporal and Spatial Boundaries**: By redefining time and space in terms of particle behavior, the Vetter Theory suggests that boundaries in time and space are not absolute but depend on proton interactions. This perspective aligns with a deterministic universe, where the structure of time and space emerges naturally from particle dynamics.

This view of time and space as particle-based phenomena offers a cohesive framework for understanding the structure of reality, removing the need for spacetime curvature and opening new avenues for exploring cosmic and atomic phenomena.

Implications for the Concept of Energy and Existence

The Vetter Theory's proton-centered framework offers a unique interpretation of energy as a product of particle interactions rather than an abstract quantity. In this model, energy is not a separate entity but an intrinsic property of particle interactions, with protons as the ultimate source of all energy. This approach redefines energy and existence, offering a new way to think about the universe's dynamics.

- **Energy as Particle Interaction Intensity**: Energy is not a distinct force or quantity; it is the intensity of interactions between particles, specifically protons and quarks. In this model, energy flows through the universe as a function of particle density and behavior, making it a product of interactions rather than a stand-alone entity.
- **Existence as Continuous Particle Dynamics**: Reality exists through the constant interactions and dynamics of particles. The Vetter Theory suggests that existence is not static but a continuous process of particle behavior, with protons and quarks driving all observable phenomena. This process-based view of existence implies that reality is inherently active and dynamic.
- **Causality as Determined by Proton and Quark Interactions**: With all forces emanating from protons, causality becomes a natural consequence of particle interactions. Events unfold predictably based on proton and quark density, aligning with a deterministic universe where causation is a straightforward result of particle behavior.

This interpretation of energy and existence as outcomes of particle interactions suggests that the universe is a self-sustaining system, where all phenomena arise from the natural dynamics of protons and quarks, providing a more integrated view of reality.

Reinterpreting Reality Through a Unified, Particle-Driven Lens

By positioning reality as a function of particle interactions, the Vetter Theory offers a cohesive view of existence that is both deterministic and structured. This perspective simplifies our understanding of time, space, and energy, suggesting that reality is a direct result of proton and quark behavior. In this particle-centered model, existence becomes predictable and measurable, grounding the universe in observable interactions and eliminating the need for abstract fields and complex forces.

- **A Foundation for Further Scientific Exploration**: The theory's emphasis on tangible, measurable particles provides a solid foundation for scientific exploration, encouraging researchers to focus on observable phenomena rather than hypothetical fields or carriers.
- **A Structured, Deterministic Reality**: By defining reality through proton dynamics, the Vetter Theory presents a structured and deterministic model of existence, positioning the universe as a system that can be fully understood through particle behavior.
- **Implications for Human Understanding of the Universe**: This particle-driven interpretation of reality challenges established frameworks, suggesting that understanding existence may be simpler and more accessible than previously thought. It encourages a shift toward a model where reality is defined by measurable events, offering a streamlined approach to science and knowledge.

This redefined view of reality, grounded in proton interactions and quark dynamics, sets the stage for a broader rethinking of the universe's structure and our place within it. By encouraging a deterministic, measurable approach, the Vetter Theory opens the door to a simplified, cohesive model of existence that challenges traditional interpretations and invites new perspectives on fundamental questions about reality.

5.2 A Deterministic Universe: Reconsidering Free Will and Uncertainty

Moving from Probabilistic Models to Determinism

The Vetter Theory, by centering on proton-driven interactions as the foundation of all forces, presents a deterministic view of the universe that challenges the probabilistic framework of quantum mechanics. Instead of relying on uncertainty and randomness in particle behavior, this model suggests that all events are governed by predictable, measurable interactions among protons and quarks. In this deterministic universe, there is no need for probabilistic interpretations, as outcomes are predetermined by particle dynamics.

- **Replacing Quantum Uncertainty with Particle Predictability**: Quantum mechanics describes particle behavior as inherently uncertain, relying on probabilities and wave functions to predict outcomes. In contrast, the Vetter Theory redefines particle interactions as predictable events driven by proton dynamics, replacing uncertainty with determinism.

- **Causality as a Function of Proton Behavior**: In this framework, causality emerges directly from proton and quark interactions. Since all forces are extensions of the strong force, every event follows naturally from preceding particle states, leading to a universe where causation is simple and deterministic.
- **Challenging the Role of Probability in Science**: By suggesting that particle interactions can be fully predicted based on proton density and behavior, the Vetter Theory challenges the role of probability in scientific models. It proposes that randomness in particle behavior is not inherent but rather a result of incomplete understanding, encouraging a shift toward deterministic explanations.

This deterministic view fundamentally alters the traditional quantum perspective, proposing a universe that is fully predictable, governed by particle interactions rather than probability.

The Role of Proton-Based Determinism in Decision-Making

The Vetter Theory's deterministic framework extends to questions of human behavior and decision-making, as it suggests that all phenomena, including consciousness and actions, are ultimately outcomes of proton and electron interactions. If every event is predetermined by particle behavior, then questions of free will, agency, and autonomy must also be reconsidered within this particle-centered model.

- **Decision-Making as a Series of Determined Interactions**: In a deterministic universe, human decisions are not random or free but are instead the result of a complex series of proton and electron interactions within the brain. This view positions consciousness and decision-making as outcomes of molecular interactions, which unfold predictably based on initial conditions.
- **Reevaluating Free Will**: The notion of free will becomes less absolute in a particle-driven universe. If all actions are determined by proton-electron dynamics, then individual choices are ultimately outcomes of predetermined interactions, challenging traditional views of autonomy and self-determination.
- **Implications for Ethics and Responsibility**: A deterministic model prompts reconsideration of concepts like responsibility and moral accountability. If actions are predetermined by particle interactions, then the basis for ethical judgment shifts, potentially encouraging a focus on understanding behavior rather than assigning blame.

By grounding decision-making in particle interactions, the Vetter Theory challenges the concept of free will, suggesting that human behavior is predictable and governed by the same deterministic forces that shape the rest of the universe.

Exploring the Boundaries of Human Knowledge

In a deterministic universe where all events are governed by proton and quark interactions, the Vetter Theory raises questions about the boundaries of human knowledge. If reality is fully predictable, it suggests that, theoretically, humans could achieve complete understanding of the universe's behavior. However, practical limitations in measuring particle interactions might still leave certain phenomena beyond our grasp, raising questions about the limits of human knowledge.

- **Predictability and the Potential for Absolute Knowledge**: Since the Vetter Theory positions all forces and interactions as predictable outcomes of proton dynamics, it theoretically allows for complete knowledge of the universe. This deterministic framework suggests that, with sufficient understanding, humans could predict events on both cosmic and atomic scales with precision.
- **The Practical Limits of Measurement**: While the theory allows for complete predictability, practical limitations in observing and measuring particle interactions could prevent humans from achieving absolute knowledge. These limitations raise questions about whether certain aspects of reality will remain inherently unknowable, even within a deterministic framework.
- **Implications for Scientific Inquiry and Discovery**: If the universe is fully predictable, then the purpose of scientific inquiry shifts from discovering new laws to refining measurements and models. This approach encourages science to focus on precision and accuracy, aiming to refine understanding of particle dynamics rather than uncovering new uncertainties.

The Vetter Theory's deterministic view suggests that absolute knowledge may be achievable in theory, yet practical limitations could mean that some mysteries remain, offering a nuanced perspective on the boundaries of human understanding.

Reconsidering Free Will and Uncertainty in a Deterministic Universe

The Vetter Theory's particle-centered model presents a deterministic view of the universe that redefines concepts of free will, uncertainty, and the limits of knowledge. By framing all events as outcomes of predictable particle interactions, it challenges traditional views of autonomy, randomness, and the unknown, offering a cohesive framework that integrates human behavior and decision-making into a larger deterministic system.

- **A Universe Without Inherent Uncertainty**: This particle-driven model eliminates the need for probabilistic explanations, suggesting that all phenomena can be predicted based on particle dynamics, removing randomness from the fabric of reality.
- **Human Agency Within a Determined Framework**: While free will may be limited, understanding behavior as a result of particle interactions offers insights into human agency, positioning individual actions as predictable outcomes of molecular dynamics.

- **Science as Precision in a Deterministic System**: With the potential for complete predictability, scientific inquiry focuses on refining measurements and understanding particle interactions rather than navigating uncertainty, encouraging a shift toward precision-based research.

By offering a deterministic, structured view of reality, the Vetter Theory invites reconsideration of foundational concepts like free will and uncertainty, suggesting a universe that is both knowable and measurable within the limits of human inquiry.

5.3 The Meaning of the Universe: Purpose, Origin, and Structure

Cosmic Origins Without a Beginning

In traditional cosmology, the universe is believed to have originated from a single event—commonly described as the Big Bang—marking the beginning of time, space, and matter. However, the Vetter Theory suggests a different perspective. By proposing a universe that continuously cycles matter and energy through particle interactions, the theory challenges the idea of a singular beginning, presenting an eternal and self-sustaining universe.

- **A Universe of Continuous Creation and Annihilation**: According to the Vetter Theory, the universe is an ongoing process where matter is constantly created and destroyed, with protons and quarks forming new particles and structures. This perspective removes the need for a singular creation event, suggesting that existence is timeless and cyclical.
- **The Implication of Infinite Existence**: In a universe with no beginning or end, questions about the origin of matter and energy are reframed. The Vetter Theory suggests that protons and quarks, through their interactions, perpetuate the cosmos indefinitely, implying that existence itself is a natural, self-sustaining state.
- **Challenging the Concept of Time as Linear**: Without a singular beginning, time may not be strictly linear but cyclical, defined by recurring particle interactions rather than a one-way progression from past to future. This idea challenges linear time models, suggesting that cosmic events are not leading toward an endpoint but are part of an eternal, self-perpetuating cycle.

This concept of an origin-less universe invites a rethinking of existence, suggesting that the universe may simply "be" without needing an origin story, offering a timeless perspective on the nature of reality.

Implications for the Concept of Purpose in the Universe

The Vetter Theory's interpretation of a self-sustaining, particle-driven universe raises questions about purpose and meaning. If the universe operates independently of any external design or intent, this deterministic model challenges traditional views that seek meaning or purpose in cosmic events, suggesting instead that the universe's structure and behavior are products of intrinsic particle interactions.

- **Purpose as Emergent Rather Than Imposed**: In a self-regulating universe, purpose is not imposed by an external force but arises naturally from particle behavior. Every cosmic phenomenon, from gravity to light, is an outcome of proton and quark interactions, making purpose an emergent property rather than a predefined goal.
- **A Universe Without Intrinsic Meaning**: If the universe is governed solely by particle dynamics, then concepts of purpose may be human constructs rather than inherent qualities of the cosmos. The Vetter Theory suggests that the universe's behavior is not directed toward any particular goal, leaving questions of purpose to human interpretation.
- **Redefining Cosmic Significance**: While the universe may lack an overarching purpose, understanding its structure as a self-sustaining system offers a form of significance. By comprehending the universe as a cohesive, particle-driven entity, humans can derive meaning from its complexity and interconnectedness, even if no higher purpose exists.

This view encourages a shift from seeking an external purpose to appreciating the universe's self-sustaining dynamics, finding meaning in the structure and behavior of reality itself.

Understanding the Universe as a Structured System

The Vetter Theory's quark matrix and proton density framework present a universe that is inherently structured, where galaxies, clusters, and cosmic filaments are organized through particle density variations. This structured system suggests that the universe is not random or chaotic, but ordered and predictable, governed by the natural behavior of particles.

- **The Cosmic Web as a Product of Quark Density**: By defining the universe's structure in terms of quark density, the Vetter Theory presents the cosmic web as an ordered system, where galaxies form along filaments and clusters according to predictable density variations. This view aligns with observations of large-scale structure, suggesting a cohesive and structured cosmos.
- **Gravitational Fields as an Outcome of Particle Clustering**: Gravity emerges naturally from proton density, creating ordered fields that shape stars, galaxies, and planetary systems. This structure-driven view implies that the universe's layout is a direct result of particle behavior, emphasizing the predictable and ordered nature of reality.
- **Order Without Design**: The structured nature of the universe, according to the Vetter Theory, does not imply external design. Instead, this order is an intrinsic quality of particle dynamics, suggesting that the universe's structure is the natural outcome of proton and quark behavior, not a product of intentional planning.

This view of the universe as a structured, particle-driven system reinforces the idea that while cosmic order exists, it arises organically, without need for an external designer or architect.

Purpose, Origin, and Structure: A New Perspective on the Universe's Meaning

The Vetter Theory's particle-centered framework presents a universe that is self-sustaining, timeless, and structured through natural particle interactions. By challenging the need for a singular origin and external purpose, it offers a unique perspective on the nature of existence, proposing that meaning in the universe may be found in its structure and predictability rather than in its origin or purpose.

- **A Timeless, Self-Sustaining Cosmos**: This model's rejection of a singular beginning suggests a universe that has always existed, challenging traditional cosmology and introducing the concept of an origin-less reality.
- **Meaning Through Structure, Not Intent**: By presenting the universe as structured and ordered through particle dynamics, the Vetter Theory offers a way to find meaning in the cosmos without requiring external purpose, inviting humans to appreciate the universe's complexity and predictability.
- **A Universe Defined by Natural Order**: The theory's structured view of reality suggests that the universe is naturally organized, where cosmic phenomena are governed by the simple, consistent rules of particle interactions, providing a coherent and predictable model of existence.

This perspective positions the Vetter Theory as a framework that redefines the meaning of the universe, offering an explanation for cosmic structure and behavior that is rooted in the natural, self-sustaining dynamics of particles.

5.4 Human Identity in a Particle-Driven Universe

The Composition of Consciousness and Identity

The Vetter Theory's proton-centered model suggests that consciousness and identity are products of molecular and atomic interactions, driven by the behavior of protons, electrons, and quarks. This particle-based framework proposes that the human mind, rather than being a separate or mysterious entity, emerges directly from the dynamics of matter, making consciousness an outcome of deterministic processes.

- **Consciousness as Emergent from Particle Interactions**: In this framework, consciousness arises from the organized interactions of protons, electrons, and quarks within the brain's molecular structure. This model suggests that conscious experience is the natural result of complex particle dynamics, positioning it as a phenomenon that can be understood within the laws of particle behavior.
- **Identity as a Continuum of Particle Configurations**: Human identity, under the Vetter Theory, is not static but a sequence of configurations of protons and electrons. Just as consciousness emerges from particle interactions, so does identity, which continually evolves as particle arrangements within the brain change in response to experiences.

- **Redefining the Self as Particle-Based**: This perspective challenges traditional notions of a fixed, independent self, suggesting instead that identity and consciousness are fluid, continually reconstructed through ongoing particle interactions. In this view, the self is as dynamic as the particles that constitute it, aligning identity with a particle-driven framework of existence.

By redefining consciousness and identity as outcomes of particle interactions, the Vetter Theory presents a cohesive view of the human experience that integrates identity with the universe's deterministic structure.

The Role of Proton Dynamics in Biological Systems

Beyond consciousness, the Vetter Theory suggests that proton dynamics play a fundamental role in all biological processes. By focusing on particle behavior, this model implies that life itself is the result of coordinated molecular interactions, with protons and quarks as the foundational units of biological organization.

- **Life as a Network of Proton-Based Interactions**: Biological processes, from cellular activity to genetic expression, can be seen as networks of proton-electron interactions. This interpretation suggests that life emerges from molecular patterns determined by proton dynamics, grounding biology in particle physics.
- **Evolution as Deterministic Particle Dynamics**: Evolutionary change, in this view, is not a random process but an outcome of predictable particle interactions within DNA and cellular structures. By positioning evolution within a deterministic framework, the Vetter Theory suggests that biological diversity and adaptation follow predictable patterns driven by proton-based molecular interactions.
- **Complexity in Biological Systems Through Proton Interactions**: The complexity of life, from multicellular organisms to ecosystems, emerges naturally from proton interactions within biological molecules. This perspective integrates biology with particle physics, suggesting that the diversity of life is a product of fundamental particle behavior.

By framing biological systems within a particle-driven model, the Vetter Theory suggests that life and evolution are consistent with the same deterministic principles that govern atomic and cosmic scales.

Free Will, Agency, and the Boundaries of Self

The Vetter Theory's deterministic universe raises questions about free will, agency, and the concept of the self. If human identity and consciousness are governed by predictable particle interactions, then the boundaries of the self are defined by molecular and atomic structures, positioning human agency as an outcome of particle dynamics rather than autonomous will.

- **Agency as Determined by Proton-Electron Interactions**: In this framework, human actions are the result of proton and electron interactions within the brain and body, making decisions predictable outcomes of molecular configurations. This view challenges traditional ideas of free will, suggesting that choices are predetermined by the behavior of particles.

- **Redefining Autonomy Within Determinism**: While free will may be limited, understanding actions as outcomes of particle interactions provides a new way to conceptualize autonomy. In this view, individuals operate within a deterministic framework, with agency emerging as a natural outcome of structured molecular interactions rather than a truly independent force.
- **The Fluid Boundaries of Selfhood**: By grounding identity in proton and quark dynamics, the Vetter Theory suggests that the self is not an isolated, fixed entity but a fluid arrangement of particles. This view implies that the boundaries of selfhood are flexible, evolving in response to ongoing molecular changes within the body and brain.

This perspective challenges traditional concepts of autonomy and self, proposing that human agency and identity are products of a deterministic, particle-driven universe.

Human Identity in a Deterministic, Particle-Centered Cosmos

By positioning consciousness, identity, and agency as outcomes of particle interactions, the Vetter Theory presents a deterministic view of human identity within the larger framework of a particle-driven universe. This perspective integrates human experience with the cosmos, suggesting that individuality and consciousness are natural outcomes of molecular and atomic dynamics.

- **Consciousness and Identity as Deterministic Phenomena**: In this view, consciousness is not mysterious or separate from physical reality but emerges naturally from particle interactions, aligning human identity with the same deterministic principles that govern the universe.
- **Life and Evolution as Outcomes of Proton Dynamics**: Biological complexity, including life and evolutionary processes, is consistent with a deterministic model, positioning the diversity of life as a predictable result of molecular interactions.
- **Agency and Self as Fluid, Deterministic Constructs**: The Vetter Theory's perspective on agency challenges the concept of a fixed self, suggesting that identity and autonomy are flexible outcomes of ongoing particle interactions rather than independent forces.

This deterministic model of human identity provides a unified view of existence, positioning consciousness, life, and agency as structured phenomena that fit seamlessly within the universe's particle-driven framework.

5.6 Embracing Change: Scientific Paradigms and Evolution

Scientific Revolutions and Paradigm Shifts

The Vetter Theory, with its particle-centered, deterministic approach, challenges established scientific frameworks and aligns with historical shifts in science where new models replaced conventional wisdom. This section examines the theory as part of a larger pattern of scientific revolutions, where breakthroughs often require rethinking the foundations of knowledge and embracing new perspectives.

- **A New Paradigm in Physics and Cosmology**: Like past revolutions—such as the shift from geocentrism to heliocentrism, or Newtonian mechanics to Einstein's relativity—the Vetter Theory proposes a radical change. By redefining forces, matter, and cosmic structure, it invites a re-evaluation of fundamental principles, positioning itself as a new paradigm in physics and cosmology.
- **Questioning Long-Held Assumptions**: The theory encourages scientists to challenge ingrained assumptions, such as the need for force carriers and spacetime curvature. This openness to questioning core beliefs is essential to scientific progress, as it allows researchers to explore alternatives that may provide simpler, more cohesive explanations.
- **The Role of Skepticism and Curiosity in Advancement**: Paradigm shifts often face initial resistance, as new ideas challenge established models. The Vetter Theory highlights the value of curiosity, skepticism, and a willingness to embrace change as essential qualities in science, fostering an environment where innovative theories can thrive.

By aligning with the history of paradigm shifts, the Vetter Theory emphasizes the importance of adaptability and openness to new ideas, positioning itself as a model for future scientific revolutions.

Integrating New Theories with Traditional Knowledge

Although the Vetter Theory presents a fundamentally different framework, it does not dismiss past discoveries or traditional knowledge. Instead, it offers opportunities for integration, where insights from established theories can be adapted to fit within a particle-centered, deterministic model. This section discusses how traditional concepts can coexist with the Vetter Theory, enriching scientific understanding through synthesis and reinterpretation.

- **Adapting Classical Concepts for Particle-Centered Models**: While the theory challenges fields like quantum mechanics and relativity, it also allows for reinterpretation. For instance, gravitational phenomena can still be observed and studied within the Vetter framework, with the theory offering an alternative explanation based on proton density rather than spacetime curvature.
- **Honoring Historical Discoveries Through New Perspectives**: The Vetter Theory respects the contributions of past discoveries, seeing them as steps toward a more cohesive understanding of reality. By building on classical concepts like force and causality, the theory integrates past knowledge into a simplified, unified framework, creating continuity with traditional science.

- **Encouraging Incremental Shifts Alongside Revolutionary Changes**: While the theory proposes a new model, it also supports gradual integration. Established theories and concepts can be adapted to fit within a particle-driven model, encouraging scientists to test, validate, and refine insights from both traditional and new perspectives.

This approach fosters a spirit of synthesis, where past and present knowledge can coexist, allowing scientists to draw from the strengths of both established and innovative models.

The Role of Curiosity and Innovation in Science

The Vetter Theory highlights the importance of curiosity, adaptability, and a mindset open to discovery as core components of scientific progress. By challenging conventional models and encouraging new ways of thinking, it reinforces the role of creativity and innovation in advancing human knowledge, positioning science as a continuous journey of exploration.

- **Curiosity as the Foundation of Discovery**: Curiosity drives scientific inquiry, pushing researchers to question assumptions, explore new ideas, and seek deeper understanding. The Vetter Theory embodies this spirit, suggesting that science thrives when curiosity is nurtured, leading to innovative discoveries that may redefine existing models.
- **Innovation Through Simplification**: The theory's emphasis on a particle-centered, deterministic model reflects the value of simplifying complex phenomena. By reducing the universe to measurable interactions of protons and quarks, the Vetter Theory encourages a pursuit of elegance and simplicity, showing that innovation often lies in finding cohesive, straightforward explanations.
- **Embracing Science as an Evolving Discipline**: Recognizing that science is never static, the Vetter Theory promotes a view of scientific knowledge as continuously evolving. Each new theory adds layers of understanding, inspiring future generations to build on current insights and further refine humanity's view of reality.

This perspective on science celebrates curiosity, resilience, and a commitment to progress, positioning the Vetter Theory as a catalyst for future innovations and a reminder of the dynamic, ever-changing nature of scientific inquiry.

A Catalyst for Future Paradigms in Science

The Vetter Theory's particle-driven framework not only offers a new perspective on fundamental forces and cosmic structure but also serves as a model for embracing scientific change. By challenging established paradigms and encouraging a spirit of curiosity, it positions itself as a catalyst for future scientific evolution, inspiring researchers to question, innovate, and explore.

- **Encouraging a Culture of Exploration and Adaptability**: The theory promotes an open-minded approach to science, where exploration and adaptability are valued as essential qualities. This mindset allows for the continuous re-evaluation of models, ensuring that science remains responsive to new ideas and discoveries.
- **A Commitment to Simplicity and Cohesion**: By seeking a unified, simplified explanation of reality, the Vetter Theory embodies the principle that science advances through cohesion. This commitment to streamlined models serves as a guiding principle for future theories, encouraging scientists to pursue elegant, cohesive frameworks.
- **Inspiring Future Generations of Thinkers and Innovators**: As a radical shift in understanding, the Vetter Theory inspires future generations to push boundaries, explore beyond established knowledge, and cultivate a mindset of discovery. It positions itself as a foundation for future scientific revolutions, inviting researchers to build on its principles and continue the journey of scientific exploration.

This final section of Chapter 5 celebrates the Vetter Theory as a model for embracing change, fostering a spirit of curiosity and innovation that drives scientific evolution and positions humanity on the path toward greater understanding.

Chapter 6

6.1 Designing Experiments to Test Key Predictions

Developing High-Precision Particle Interaction Experiments
The Vetter Theory's central claim—that all forces originate from proton interactions and that mass arises directly from proton-electron dynamics—suggests a need for high-precision particle interaction studies. By observing these interactions in controlled settings, researchers can test the theory's predictions regarding mass, force manifestation, and the deterministic nature of particle behavior.

- **Mass Generation from Proton Density**: Current models in particle physics attribute mass to the Higgs field, but the Vetter Theory posits that mass results from proton density and electron interactions. Particle accelerators like CERN's Large Hadron Collider (LHC) or Fermilab's Tevatron could be adapted to investigate this hypothesis by tracking proton-electron interactions under various conditions, focusing on mass fluctuations in dense proton clusters.
- **Testing Force Manifestations at Subatomic Scales**: The theory suggests that all forces are extensions of the strong force, manifesting differently depending on particle configurations. Experiments designed to isolate proton interactions, particularly at high densities, could reveal new insights into force behavior, potentially validating the theory's claims about unified force dynamics.

- **Deterministic Patterns in Particle Decay and Transformation**: By observing the behavior of particles in decay processes, researchers can test the theory's deterministic model. If particle interactions are truly predictable, experimental results should show consistent decay patterns, particularly in processes like beta decay. High-precision detectors can capture these decay sequences, offering data to support or challenge the theory's deterministic framework.

These experiments aim to capture the underlying dynamics of particle behavior, providing measurable evidence for the theory's claims about mass and force unification through proton interactions.

Observing Gravitational and Electromagnetic Effects Around Dense Objects

One of the Vetter Theory's unique predictions is that gravitational and electromagnetic effects around dense proton clusters will differ from those predicted by traditional models. By focusing on gravitational lensing, waveforms, and electromagnetic behavior near dense objects, researchers can gather observational evidence to test the theory's particle-driven model of gravity and light.

- **Gravitational Lensing Patterns Around Dense Objects**: According to the theory, gravitational lensing should result from electron-photon interactions near dense proton clusters rather than spacetime curvature. Observing asymmetrical or intensity-dependent lensing patterns around dense objects like black holes and neutron stars could reveal these particle-based effects, providing data to compare with traditional curvature-based models.
- **Waveform Analysis in Gravitational Wave Observations**: Traditional black hole mergers are expected to produce specific gravitational waveforms based on general relativity. However, if the Vetter Theory's interpretation holds, waveforms from mergers involving proton-dense black holes would display unique signatures reflecting proton density rather than singularity-driven curves. Advanced gravitational wave detectors, like LIGO and Virgo, could capture these distinct waveforms, providing an opportunity to directly test the theory's predictions.
- **Electromagnetic Emissions and High-Energy Radiation**: The theory suggests that high-energy radiation near black holes, quasars, and dense stars results from electron ejections interacting with proton clusters. X-ray and gamma-ray observatories, such as NASA's Chandra X-ray Observatory and the European Space Agency's XMM-Newton, could observe these emissions to determine if they align with the particle-driven model. If electromagnetic waves behave according to proton interactions rather than field-based forces, their behavior near these objects will show measurable differences.

These observational experiments could provide crucial evidence to support or challenge the Vetter Theory's claims, particularly regarding gravity and electromagnetic interactions around dense objects.

Mapping Quark Density in Cosmic Structures
The Vetter Theory posits that quarks form a matrix that fills intergalactic space, creating gravitational effects traditionally attributed to dark matter. By mapping quark density within galaxies and galactic clusters, researchers can test this model, aiming to observe whether quark distribution aligns with gravitational effects currently associated with dark matter.

- **Galaxy Rotation Curves and Quark Distribution**: The theory suggests that quark density in galaxies should correlate with rotation curves, eliminating the need for dark matter halos. Astronomers could use radio and optical telescopes, such as the Atacama Large Millimeter/submillimeter Array (ALMA) or the Very Large Telescope (VLT), to observe rotation patterns in galaxy clusters, testing for quark density-based gravitational effects.
- **Quark-Based Gravitational Lensing**: By mapping quark distribution in galaxy clusters, researchers could examine gravitational lensing patterns to see if they align with quark density variations rather than hypothetical dark matter particles. Instruments like the Hubble Space Telescope and the James Webb Space Telescope (JWST) could be used to capture these lensing effects in regions where dark matter has been hypothesized, providing observational evidence for the theory's quark matrix model.
- **Large-Scale Structure and Cosmic Web Formation**: The Vetter Theory predicts that cosmic web structures—filaments, voids, and galaxy clusters—are influenced by quark density. Surveys like the Sloan Digital Sky Survey (SDSS) and the Dark Energy Survey (DES) could measure galaxy distribution to determine if the cosmic web aligns with predicted quark density variations, offering another way to test the quark-based explanation for dark matter.

By mapping quark density and comparing it with gravitational observations, researchers could determine whether quarks provide a viable explanation for dark matter effects, potentially validating a central claim of the Vetter Theory.

A New Era of Precision-Based Experiments and Observations
The Vetter Theory provides a roadmap for precision-based experiments and observations that focus on measurable, particle-driven phenomena. By designing experiments that capture the theory's unique predictions about mass, gravity, and dark matter, scientists can test its viability, contributing valuable data to the ongoing exploration of particle-centered models.

- **Refining Experimental Techniques for Particle Interactions**: As particle accelerators and detectors adapt to study proton interactions in greater detail, scientists have the opportunity to validate or challenge the theory's claims about mass and force unification.

- **Pushing the Boundaries of Gravitational and Electromagnetic Observations**: Gravitational wave and high-energy observatories can capture new insights into the behavior of dense objects, offering potential support for a particle-driven model of gravity and light.
- **Mapping Quark Density as a Dark Matter Alternative**: By exploring quark density within galaxies and clusters, researchers can test the theory's quark-based explanation for dark matter, providing a new lens through which to interpret galactic and cosmic structure.

This section outlines practical approaches for testing the Vetter Theory's predictions, establishing a foundation for experimental and observational research that may validate its particle-driven model of the universe.

6.2 Leveraging Observational Astronomy for Real-World Validation

Testing Cosmic Background Radiation (CBR) Predictions

The Vetter Theory reinterprets cosmic background radiation (CBR) as an ongoing, quark-based radiation rather than as a remnant of the Big Bang. According to this model, localized variations in quark density should cause minor fluctuations in CBR intensity. By examining these variations, astronomers can test whether CBR aligns with the theory's particle-centered explanation.

- **Localized CBR Variations and Quark Density**: The Vetter Theory suggests that CBR intensity will fluctuate according to local quark density rather than displaying uniformity. Observing these localized differences could help determine if CBR is indeed linked to quark interactions. Using high-sensitivity instruments like the Planck Satellite and upcoming missions, astronomers could analyze minute variations in CBR across different regions, seeking correlations with predicted quark densities.
- **Testing for Ongoing Quark-Based Radiation Patterns**: If CBR results from continuous quark interactions, its distribution should differ from the uniformity expected under the Big Bang model. High-precision surveys of CBR intensity could reveal subtle patterns that align with a particle-driven origin, potentially validating the theory's reinterpretation of cosmic background radiation.
- **Long-Term Observations for Dynamic CBR Fluctuations**: If CBR is a result of ongoing quark interactions, astronomers could conduct long-term studies to observe dynamic fluctuations over time. Instruments like the South Pole Telescope (SPT) could monitor CBR intensity, seeking variations that support a quark-driven radiation model.

These observations provide an opportunity to challenge or validate the Big Bang interpretation of CBR, testing whether quark-based radiation can explain cosmic background patterns.

Exploring Light Behavior in High-Density Regions

According to the Vetter Theory, gravitational lensing is not caused by spacetime curvature but results from electron-photon interactions near dense proton clusters. Observing light behavior around massive objects like black holes and neutron stars could reveal particle-based lensing effects distinct from those predicted by general relativity.

- **Intensity-Dependent Lensing Patterns**: If gravitational lensing results from electron-photon interactions, then lensing patterns should show intensity-based variations rather than uniform curvature. High-resolution telescopes like the Event Horizon Telescope (EHT) and JWST could capture light bending around black holes and neutron stars, allowing astronomers to test whether lensing aligns with the Vetter Theory's predictions.
- **Asymmetrical Light Bending in Non-Spherical Objects**: The theory suggests that lensing should vary based on the shape and density of the proton clusters. Observing asymmetrical lensing patterns around dense, irregularly shaped objects—such as neutron stars—could reveal these effects, contrasting with general relativity's expectations of symmetric spacetime curvature.
- **High-Frequency Light and Quark Interactions**: The theory predicts that high-frequency light, such as X-rays and gamma rays, may interact differently around dense objects due to proton and electron dynamics. By observing the behavior of high-energy light near neutron stars or active galactic nuclei (AGN), astronomers could determine if particle interactions drive these unique lensing effects, providing evidence for the theory.

Observing light behavior in high-density regions offers a way to test the Vetter Theory's claims about gravitational lensing and electron-photon interactions, potentially providing a new perspective on the bending of light around massive objects.

Localized Expansion Rate Observations

The Vetter Theory posits that cosmic expansion is driven by quark pressure, which varies depending on quark density across regions. This model predicts localized differences in expansion rates, suggesting that regions with varying quark density should exhibit slight variations in their rate of expansion. By examining these regional differences, astronomers can test the theory's interpretation of dark energy as quark-based pressure.

- **Measuring Expansion Rate Variations Across Regions**: Using data from large-scale surveys like the Dark Energy Survey (DES) and the Euclid mission, astronomers can measure expansion rates across different regions of the universe. By correlating these rates with quark density predictions, they can determine if quark pressure influences localized expansion in ways consistent with the Vetter Theory.

- **Galaxy Cluster Distribution and Regional Expansion Effects**: Since quark density should affect both cosmic expansion and gravitational attraction, galaxy clusters should exhibit slight differences in spacing and movement patterns based on local quark pressure. Observing the distribution and dynamics of galaxy clusters in high- and low-quark-density regions could reveal correlations between quark pressure and expansion effects.
- **Testing for Quark-Driven Expansion Patterns**: The Vetter Theory suggests that localized expansion variations should align with quark density rather than appearing uniformly. By analyzing the structure and growth rate of cosmic voids and filaments, researchers can test whether these variations follow a pattern consistent with quark-driven expansion, challenging or supporting the dark energy model.

By examining regional expansion rates, astronomers can test the Vetter Theory's view of quark-based cosmic expansion, offering an alternative to the current dark energy interpretation.

Leveraging Observational Astronomy to Validate the Vetter Theory

The Vetter Theory's unique predictions about CBR, light behavior, and cosmic expansion offer a roadmap for observational astronomy, guiding astronomers to seek measurable data that could support or refute its claims. By focusing on particle-based explanations, researchers can use advanced observational tools to directly test the theory's particle-centered model.

- **Precision Observations of CBR and Quark Density**: Observing localized variations in cosmic background radiation provides an opportunity to test the theory's quark-driven radiation model, potentially challenging the traditional Big Bang interpretation.
- **High-Resolution Lensing Studies in Dense Regions**: Observing lensing patterns around dense objects can reveal electron-photon interactions, providing an alternative view of gravitational lensing that aligns with the Vetter Theory's particle-based explanation.
- **Localized Expansion Rate Analysis for Dark Energy Alternatives**: Examining regional variations in expansion rates allows astronomers to explore quark-based pressure as an explanation for cosmic expansion, offering a testable alternative to the dark energy model.

This approach to observational astronomy provides practical methods for testing the Vetter Theory, enabling scientists to gather real-world data that can validate its unique particle-centered model of the universe.

6.3 Developing Interdisciplinary Collaboration Models

Uniting Particle Physics and Cosmology for Dark Matter Research

The Vetter Theory's quark-based model of dark matter presents an opportunity for collaboration between particle physicists and cosmologists. By combining expertise in quark interactions and galactic structure, researchers can examine the role of quarks in cosmic phenomena, potentially offering an alternative explanation for dark matter.

- **Joint Experiments on Quark Density in Galactic Clusters**: Particle physicists and cosmologists could work together to study quark density in galactic clusters, comparing observed gravitational effects with predicted quark density patterns. Combining high-precision particle detectors with large-scale cosmological surveys, such as SDSS, could reveal insights into quark-based gravitational influences.
- **Integrated Simulations for Quark Matrix Dynamics**: Creating simulations that combine particle physics models with cosmic structure formation could allow researchers to visualize how quark density shapes galaxies and clusters. By integrating quark behavior into cosmological simulations, scientists could observe how the quark matrix influences large-scale structures and dark matter effects.
- **Collaborative Observational Studies of Galaxy Rotation Curves**: Studying galaxy rotation curves using particle physics insights could offer a new perspective on gravitational behavior. Teams could use observatories like ALMA and VLT to observe galactic rotation while particle physicists provide data on quark interactions, exploring whether quark density explains the effects traditionally attributed to dark matter.

This interdisciplinary approach to dark matter research enables scientists to explore a unified model, combining insights from particle physics and cosmology to test the Vetter Theory's quark-based predictions.

Biological Applications and Consciousness Studies

The Vetter Theory's suggestion that proton interactions influence biological systems, including consciousness, provides a foundation for collaboration between particle physicists, molecular biologists, and neuroscientists. By examining the role of proton dynamics in biological processes, researchers could explore the potential for a particle-centered understanding of life and consciousness.

- **Studying Proton Interactions in Molecular Biology**: Molecular biologists and particle physicists could jointly study the role of proton interactions in cellular processes. By examining proton-electron dynamics in biological molecules, researchers could gain insights into how particle behavior influences genetic expression, cellular function, and biochemical reactions.
- **Exploring Proton-Based Mechanisms in Neuroscience**: Neuroscientists could work with particle physicists to study how proton interactions affect neural activity, potentially linking particle behavior to cognitive processes. Using advanced imaging techniques like fMRI and positron emission tomography (PET), researchers could investigate how proton dynamics correlate with brain function, offering a new approach to understanding consciousness.

- **Developing a Particle-Centered Model of Consciousness**: By integrating insights from particle physics and neuroscience, researchers could develop a particle-centered model of consciousness. This model would explore how proton interactions create the conditions necessary for conscious experience, positioning consciousness as an emergent property of atomic and molecular dynamics rather than a purely biological phenomenon.

This interdisciplinary collaboration could pave the way for a unified approach to studying life and consciousness, combining molecular biology, neuroscience, and particle physics under the principles of the Vetter Theory.

Creating a Unified Research Framework Across Scales

One of the Vetter Theory's strengths is its applicability across atomic, biological, and cosmic scales. By establishing a unified research framework that spans these areas, interdisciplinary teams can test the theory's principles consistently, creating a comprehensive model that applies the same particle-driven explanations to phenomena on all scales.

- **Establishing Research Centers Focused on Unified Models**: Universities and research institutions could create centers dedicated to testing and expanding the Vetter Theory, bringing together experts from physics, biology, and astronomy. These centers would provide resources and collaborative space to explore the theory's predictions across disciplines.
- **Funding Programs to Support Cross-Disciplinary Studies**: Grant agencies and scientific organizations could fund interdisciplinary projects that examine the Vetter Theory's predictions across scales. By supporting cross-disciplinary research, these programs encourage collaboration, enabling scientists to test the theory's particle-centered principles in both laboratory and observational settings.
- **Developing Shared Methodologies and Data Protocols**: Establishing standardized methodologies and data-sharing protocols among researchers in different fields can streamline interdisciplinary collaboration. By creating shared protocols, scientists across disciplines can more easily compare findings and build on each other's work, enhancing the accuracy and consistency of research that tests the Vetter Theory's predictions.

This unified research framework fosters a cohesive approach to scientific inquiry, allowing researchers to explore the Vetter Theory's principles across atomic, biological, and cosmic scales, ultimately building a comprehensive model of reality.

Interdisciplinary Collaboration as a Path Forward

The Vetter Theory's particle-centered model provides a foundation for interdisciplinary collaboration, encouraging scientists from various fields to work together to test its predictions. By combining expertise in particle physics, biology, and cosmology, researchers can explore how proton and quark dynamics influence phenomena across scales, fostering a unified approach to scientific discovery.

- **A New Model for Integrated Scientific Inquiry**: By creating interdisciplinary teams, the Vetter Theory encourages a model of scientific inquiry that crosses traditional boundaries, uniting fields in a shared pursuit of knowledge and providing a holistic perspective on particle-driven phenomena.
- **A Platform for Innovative Research on Life and Consciousness**: Collaboration between physicists and biologists enables groundbreaking studies on the role of particle interactions in biological systems and consciousness, potentially leading to a new understanding of life that bridges physics and biology.
- **A Unified Framework for Testing Cosmic and Atomic Principles**: This interdisciplinary model allows scientists to apply the Vetter Theory consistently across disciplines, offering a cohesive framework for exploring cosmic and atomic phenomena through particle interactions.

By promoting collaboration across scientific fields, the Vetter Theory positions itself as a model for unified inquiry, encouraging researchers to build a comprehensive understanding of reality that is grounded in particle dynamics.

6.4 Potential Technological Applications Based on the Vetter Theory

Proton-Centered Energy Generation and Storage

The Vetter Theory's emphasis on proton interactions as the source of all forces suggests new avenues for sustainable energy generation and storage. By leveraging proton-driven dynamics, energy systems can achieve higher efficiency and sustainability, offering alternatives to traditional fuel sources.

- **Fusion Reimagined Through Proton-Electron Fusion**: Traditional fusion research focuses on combining atomic nuclei to release energy, but the Vetter Theory proposes that energy can be harnessed more efficiently through controlled proton-electron fusion. Developing this approach could lead to compact, high-output fusion reactors that provide consistent energy without producing long-lived radioactive waste.
- **Advanced Fuel Cells and Proton-Based Energy Storage**: By focusing on proton interactions, fuel cell technologies could become more efficient and stable, potentially using controlled proton density adjustments to regulate energy output. These advanced fuel cells could power everything from small electronics to large infrastructure, offering a flexible and clean energy solution.
- **Energy Systems with Minimal Heat Loss**: Since proton interactions in the Vetter Theory model are seen as highly efficient, energy systems based on this framework could significantly reduce heat loss, making them more effective for use in high-demand applications such as industrial power grids and spacecraft propulsion.

These energy solutions leverage the theory's principles to create efficient, sustainable energy sources, reducing reliance on traditional fuels and increasing the practicality of proton-centered systems.

Advances in Material Science and Electromagnetic Manipulation

The Vetter Theory's interpretation of forces as expressions of proton interactions offers new possibilities for materials that respond dynamically to environmental changes. These materials could enhance resilience, adaptiveness, and efficiency in fields ranging from electronics to construction.

- **Electromagnetically Adaptive Materials**: By tuning proton density within materials, scientists could create structures that respond to electromagnetic fields in real-time, changing properties based on external stimuli. These materials could be used in applications like smart fabrics, adaptable construction materials, and dynamic electronics.
- **Radiation-Resistant and High-Durability Components**: Proton-based materials could improve resistance to radiation, high temperatures, and physical stress. This would be particularly valuable for applications in aerospace, nuclear reactors, and other extreme environments where durability is essential for safe and consistent operation.
- **Proton-Driven Superconductors**: The Vetter Theory suggests that materials optimized for proton interactions could achieve superconductivity at higher temperatures. By designing materials that facilitate proton density adjustments, scientists could create superconductors that operate more effectively, potentially revolutionizing energy transmission, magnetic levitation, and medical imaging technologies.

These advances in material science create new opportunities for high-performance materials that are durable, adaptable, and energy-efficient, aligning with the Vetter Theory's particle-centered approach.

Space Exploration Technologies and Propulsion

The Vetter Theory's deterministic, particle-based model introduces a new framework for space exploration, suggesting ways to use proton interactions to power propulsion and improve spacecraft durability. By designing systems that operate efficiently in extreme environments, researchers could enable new exploration capabilities.

- **Proton-Based Ion Propulsion**: Traditional ion propulsion systems rely on accelerating ions to produce thrust, but a proton-centered model could increase thrust-to-power ratios, offering greater propulsion efficiency with less fuel. This approach could be particularly valuable for long-duration missions, allowing spacecraft to travel farther and faster.
- **Energy-Efficient Power for Spacecraft**: With proton-centered energy generation, spacecraft could utilize consistent energy sources without relying solely on solar power. This would allow for greater operational flexibility, enabling missions to explore distant regions of space where solar energy is insufficient.

- **Durable Materials for Extreme Conditions**: The radiation-resistant and high-durability materials enabled by proton interactions could be essential for protecting spacecraft and instruments in space. These materials would increase the resilience of space equipment, ensuring that it performs well under high-radiation conditions, temperature fluctuations, and physical stress.

These advancements in space exploration technology reflect the Vetter Theory's principles, enabling more sustainable, high-performance systems that support humanity's expansion into deeper space.

Potential Impact on Computing and Data Processing

By focusing on direct particle interactions and eliminating complex intermediary forces, the Vetter Theory opens up possibilities for faster, more efficient computing systems. These advances could revolutionize fields such as quantum computing, telecommunications, and artificial intelligence, where precision and speed are essential.

- **Proton-Based Quantum Computing Models**: The Vetter Theory's deterministic approach to particle interactions could streamline quantum computing processes, offering more reliable and accessible models for data processing. Proton-driven quantum systems could potentially perform calculations faster and with greater accuracy, enhancing data security and computational efficiency.
- **High-Speed Data Transmission**: The theory's reinterpretation of electromagnetic waves as proton-driven interactions could reduce interference and signal degradation, making data transmission faster and more stable. This could improve communication systems in both wireless and fiber-optic networks, enhancing speed and reliability.
- **Energy-Efficient Processing Units**: With a focus on direct proton interactions, computing processors could perform calculations with less energy loss, reducing heat output and power consumption. This improvement could make high-performance processors more sustainable, especially in data centers and AI research applications.

These applications in computing align with the Vetter Theory's principles, promoting high-speed, low-loss data processing that can support the future demands of information technology.

A New Era of Technology Inspired by Particle-Centered Principles

The Vetter Theory's particle-driven framework provides a foundation for a new generation of technologies that are efficient, durable, and responsive. By focusing on proton interactions and quark density, these applications enhance sustainability and performance, positioning the theory as a catalyst for technological advancement.

- **Energy and Power Generation Technologies**: Proton-based energy generation and storage systems could reduce reliance on traditional fuels, providing cleaner and more efficient power solutions that are adaptable to various scales and applications.

- **Resilient Materials for High-Performance Environments**: Advances in material science, driven by proton interactions, create resilient, adaptive materials that can withstand extreme conditions, expanding possibilities in construction, electronics, and aerospace.
- **Enhanced Computing and Telecommunications**: By simplifying data processing through proton-driven systems, the Vetter Theory offers a path to faster, more efficient computing technologies, supporting advancements in AI, telecommunications, and quantum computing.

This section demonstrates the theory's potential to inspire a new era of technology that reflects its core principles, offering practical solutions that can transform energy, materials, space exploration, and computing.

6.5 Developing Educational and Research Frameworks for Future Generations

Educational Curricula in Particle-Centered Models

To foster understanding of the Vetter Theory and its implications, educational institutions could incorporate particle-centered models into physics, cosmology, and biology courses. By introducing students to a unified, proton-driven perspective, educators can encourage young scientists to explore alternative models and approach traditional topics from new angles.

- **Introducing Particle Dynamics in General Physics**: At the undergraduate level, physics courses could begin teaching the fundamentals of the Vetter Theory, positioning proton and quark interactions as central to understanding forces and mass. This introduction would provide students with a particle-centered framework early in their studies, allowing them to approach more advanced topics with a cohesive perspective.
- **Expanding Cosmology and Astrophysics Curricula**: By including the Vetter Theory's interpretation of gravitational lensing, cosmic background radiation, and cosmic expansion, cosmology courses can present an alternative to the field-based and relativity-driven models. This approach encourages students to think critically about dark matter, dark energy, and cosmic structure through the lens of quark density and particle-driven gravity.
- **Biology and Neuroscience Courses on Proton Interactions**: In molecular biology and neuroscience, students could learn about how proton-electron dynamics influence cellular processes and cognitive functions, integrating the Vetter Theory into life sciences. This curriculum would introduce a unified approach to biological systems, emphasizing the role of particle dynamics in evolution, genetics, and consciousness.

By integrating these principles into physics, cosmology, and biology curricula, educational institutions can cultivate a generation of scientists who are well-versed in the Vetter Theory's particle-centered approach, equipping them to apply these principles in various fields.

Funding and Institutional Support for Theory-Based Research
Supporting research that tests and expands the Vetter Theory requires a commitment from institutions and funding organizations. By establishing grants, research centers, and collaborative programs, institutions can encourage scientists to explore the theory's predictions and its potential to advance scientific understanding.
- **Dedicated Grants for Particle-Centered Research**: Funding organizations could create grant programs specifically for research that tests the Vetter Theory's principles. These grants would support projects exploring proton interactions, quark density in cosmic structures, and applications in energy and materials science, providing resources for experimental validation and interdisciplinary studies.
- **Establishing Vetter Theory Research Centers**: Research institutions and universities could create specialized centers focused on testing and expanding the Vetter Theory. These centers would bring together physicists, biologists, cosmologists, and engineers to conduct studies and experiments that explore the theory's applications, fostering collaboration and innovation across disciplines.
- **Encouraging Public and Private Research Partnerships**: Collaborative programs that link academic research with private industry could support technological applications inspired by the Vetter Theory. By partnering with companies in energy, aerospace, and computing, institutions can encourage the development of technologies rooted in proton-centered models, translating theory into practical solutions.

Institutional and funding support enables scientists to rigorously test the Vetter Theory and apply its principles, providing resources that advance both experimental and theoretical research.

Inspiring Future Generations of Scientists
The Vetter Theory encourages a spirit of curiosity, exploration, and open-mindedness—qualities essential for scientific advancement. By promoting these values, educational and research programs can inspire future scientists to embrace alternative models, question established frameworks, and pursue innovation in their fields.
- **Creating Outreach Programs and Public Science Initiatives**: To inspire students and the public, institutions could develop outreach programs that introduce the Vetter Theory's concepts in accessible ways. These programs could include lectures, workshops, and interactive events that showcase the theory's particle-driven approach and its implications for understanding the universe.
- **Promoting Interdisciplinary Research in Undergraduate and Graduate Programs**: By integrating interdisciplinary research projects into educational programs, institutions can encourage students to explore particle-centered models across various fields, from physics to biology and engineering. This experience prepares students to think beyond traditional boundaries and pursue innovative research that bridges scientific disciplines.

- **Cultivating a Culture of Curiosity and Adaptability**: Emphasizing the importance of curiosity and adaptability in science helps students approach challenges with an open mind, encouraging them to explore and test new theories. By fostering a supportive environment for questioning established ideas, educators can inspire students to become future leaders in scientific discovery, guided by the principles of the Vetter Theory.

Through outreach, interdisciplinary projects, and a culture of exploration, educational institutions can inspire the next generation to advance the Vetter Theory's principles and develop new, innovative models of understanding.

A Foundation for Continued Exploration and Discovery

By integrating the Vetter Theory into educational frameworks and research programs, institutions can establish a foundation for continued exploration. This approach provides students and scientists with the tools and resources to test, expand, and apply the theory's particle-centered model, positioning it as a lasting influence on scientific inquiry.

- **A Cohesive Curriculum for Understanding Reality**: By building the Vetter Theory into curricula across physics, cosmology, and biology, educational programs can offer students a cohesive framework for understanding forces, matter, and biological systems, preparing them to apply particle-centered principles in diverse scientific contexts.
- **Institutional Support for Long-Term Research Goals**: Through grants, research centers, and collaborations with private industry, institutions can provide the support needed to pursue long-term research goals that test the Vetter Theory's predictions and potential applications.
- **Inspiring a Future of Interdisciplinary Science**: By fostering a culture of curiosity, adaptability, and collaboration, educational and research programs can inspire scientists to pursue interdisciplinary studies, bridging gaps between fields and exploring unified models of reality.

This educational and institutional framework encourages a future where the Vetter Theory's principles are explored, validated, and expanded upon, guiding scientific inquiry and technological innovation for generations to come.

6.6 Conclusion: Embracing a Future of Precision, Unity, and Discovery

A Call to Action for Precision in Science

The Vetter Theory's particle-centered framework challenges traditional scientific models, offering a simplified, cohesive explanation for fundamental forces, mass, and cosmic structure. By focusing on direct, measurable interactions among protons, quarks, and electrons, the theory encourages a renewed commitment to precision in science, urging researchers to refine their techniques and measurements to reveal the underlying patterns of reality.

- **Advancing Knowledge Through High-Precision Observations**: Precision is essential to validating the theory's predictions about mass, gravity, and dark matter. By focusing on accurate measurements of proton and quark interactions, scientists can gain insights into particle behavior and test the theory's deterministic framework.
- **Refining Experimental Techniques Across Disciplines**: Whether in particle physics, cosmology, or biology, precision-based experiments can help reveal the true nature of forces and matter. By developing high-sensitivity instruments and innovative data collection methods, researchers can gather detailed evidence that supports or challenges the Vetter Theory's principles.
- **A Science Built on Measurable Interactions**: The theory's emphasis on measurable particle interactions reinforces the value of empirical data, suggesting that scientific progress can be achieved by focusing on observable phenomena and grounding knowledge in measurable reality rather than abstract concepts.

By committing to precision in their work, scientists can uncover the insights needed to validate or refine the Vetter Theory's predictions, paving the way for a deeper understanding of the universe.

Fostering a Unified, Interdisciplinary Approach

The Vetter Theory's applicability across atomic, biological, and cosmic scales promotes a unified approach to scientific inquiry, encouraging researchers from different fields to collaborate and explore shared principles. This interdisciplinary framework aligns with the theory's goal of simplifying complex phenomena, emphasizing the interconnectedness of all aspects of reality.

- **Bridging Scientific Disciplines Through Shared Goals**: A particle-centered approach encourages interdisciplinary teams to work together, using common principles to explore diverse phenomena. By fostering collaboration among physicists, biologists, and cosmologists, the theory positions itself as a foundation for unified scientific exploration.
- **Expanding Research Horizons Across Scales**: The Vetter Theory's applicability to both microscopic and cosmic phenomena allows researchers to apply its principles consistently across fields, promoting a cohesive understanding of the universe. This approach encourages scientists to see connections between disciplines, inspiring innovative solutions that benefit from insights across scales.
- **Encouraging Collaborative Innovation in Science and Technology**: The theory's unified framework for understanding matter and energy opens doors for technological advancements that bridge fields, from energy generation to space exploration. By encouraging interdisciplinary research, scientists can develop practical applications that reflect the theory's principles and advance technological innovation.

This unified, interdisciplinary approach creates a cohesive framework for scientific inquiry, encouraging collaboration and the development of innovative solutions that reflect the Vetter Theory's particle-centered perspective.

The Promise of a Particle-Centered Model

The Vetter Theory's particle-driven model offers a promising path for advancing our understanding of the universe. By simplifying the nature of forces and emphasizing deterministic particle interactions, it presents a clear, measurable framework that can guide scientific inquiry for generations to come. This model encourages scientists to seek elegant, unified explanations that bring cohesion to our understanding of reality.

- **A Simplified, Cohesive Model of Reality**: The theory's focus on proton interactions as the source of all forces presents a streamlined framework, offering a more cohesive explanation for complex phenomena. This simplicity makes it easier to apply the theory across disciplines, enhancing our ability to predict and understand natural events.
- **Inspiring Future Generations to Pursue Knowledge**: As an alternative to traditional models, the Vetter Theory provides a fresh perspective that inspires curiosity and exploration. By encouraging young scientists to question established frameworks and explore new models, it fosters a culture of innovation and adaptability in scientific education.
- **A Legacy of Precision and Unity in Science**: By emphasizing measurable interactions, the Vetter Theory creates a lasting impact on scientific inquiry, encouraging a commitment to accuracy, unity, and curiosity. This legacy positions the theory as a guiding principle for future scientific revolutions, inspiring ongoing efforts to understand the universe's fundamental nature.

This particle-centered model provides a foundation for continued exploration and discovery, offering a path for scientists to pursue knowledge through a unified, cohesive approach that reflects the simplicity and elegance of reality.

Embracing a Future of Discovery and Understanding

The Vetter Theory challenges traditional scientific paradigms, encouraging a future where precision, unity, and discovery guide scientific inquiry. By emphasizing measurable particle interactions and interdisciplinary collaboration, it positions itself as a model for advancing knowledge, inspiring scientists, educators, and institutions to embrace a cohesive understanding of the universe.

- **A Science Rooted in Curiosity and Innovation**: The theory promotes a science that values curiosity, simplicity, and open-mindedness, encouraging researchers to question assumptions and pursue innovative solutions that reflect the interconnectedness of natural phenomena.

- **A Path Toward Cohesive Scientific Models**: By bridging fields and fostering collaboration, the Vetter Theory offers a framework that aligns diverse disciplines, allowing researchers to develop cohesive models that provide consistent, elegant explanations across scales.
- **A Lasting Influence on Future Scientific Inquiry**: As an invitation to explore and innovate, the Vetter Theory serves as a guiding influence for future scientific advancements, encouraging a continued commitment to understanding the universe's fundamental structure and the principles that govern it.

With this conclusion, the Vetter Theory invites scientists to embark on a journey of exploration and discovery, embracing a future where precision, unity, and curiosity illuminate the path toward a deeper understanding of reality.

www.ingramcontent.com/pod-product-compliance
Lightning Source LLC
Chambersburg PA
CBHW031438210526
45464CB00005B/2251